Scented rice (*Oryza sativa* L.) Cultivars of India: A Perspective on Quality and Diversity

Altafhusain Nadaf • Sarika Mathure
Narendra Jawali

Scented rice (*Oryza sativa* L.) Cultivars of India: A Perspective on Quality and Diversity

Springer

Altafhusain Nadaf
Department of Botany
Savitribai Phule Pune University
Pune, Maharashtra, India

Sarika Mathure
Department of Botany
Savitribai Phule Pune University
Pune, Maharashtra, India

Narendra Jawali
Molecular Biology Division
Bhabha Atomic Research Centre
Mumbai, Maharashtra, India

ISBN 978-81-322-2663-5 ISBN 978-81-322-2665-9 (eBook)
DOI 10.1007/978-81-322-2665-9

Library of Congress Control Number: 2015958799

Springer New Delhi Heidelberg New York Dordrecht London

Printed on acid-free paper

Springer (India) Pvt. Ltd. is part of Springer Science+Business Media (www.springer.com)

Foreword

Scented rice is attractive to humans because of low odor threshold levels of volatile compounds including 2-acetyl-1-pyrroline. This trait has been selected by humans despite it apparently reducing the resistance of the rice plant to stress and being a genetically recessive trait because it is so attractive to rice consumers. This is a great example of human selection for an agricultural trait that would be deleterious for the plant in the wild growing under natural selection but can be highly successful under human selection in a domesticated crop. Scented rice is a traditional food of India. Scented rice is of great cultural and economic significance in India.

During the past decade, Dr. Altafhusain Nadaf and his team undertook a very demanding challenge to understand the aroma volatiles contributing in rice aroma for Indian scented rice genotypes, the outcome of which is produced in the form of this book. This book provides an important and comprehensive account of the majority of basmati and non-basmati cultivars of scented rice found in India, their quality parameters, the volatiles contributing in aroma and their method of detection, diversity analysis, and the molecular markers associated with quality traits. Knowledge of these cultivars is central to the continuation of the production and use of this important food crop. I am convinced that this book will serve people with a broad spectrum of interests and will stimulate future research in this economically important category of crop.

University of Queensland
St Lucia, QLD, 4072, Australia

Prof. Robert Henry

Preface

Ancient India is one of the oldest regions where cultivation of *Oryza sativa* L. began. Among the several cultivated rice varieties, scented rice constitutes a small but special group of rice, considered as of best quality. Indian subcontinent flourishes with hundreds of indigenous aromatic cultivars and landraces, and the diversity of scented rice of India is highest in the world. The foothill of the Himalayas is the center of diversity of scented rice (Group V); from here by westward route, scented rice cultivars are distributed in different states of India. Years of natural and human selection led to the development of several locally adapted scented rice genotypes. Basmati rice – the scented pearl – is nature's gift exclusive to the Indian subcontinent. In addition to basmati varieties, many indigenous non-basmati scented rice varieties are also locally cultivated. Majority of the indigenous scented rice cultivars are small and medium grained. Though the non-basmati types do not possess all the characteristics of basmati, these varieties excel equally as far as aroma and other characters are concerned. Hence, non-basmati varieties are traded popularly in market and fetch high premium in national as well as international market. Flavor volatiles or aroma and texture are the principle sensory qualities of rice and have been rated as the major criteria for preference. More than 100 volatiles contribute in the pleasant rice aroma. Among these, 2-acetyl-1-pyrroline (2AP) possesses low odor threshold value and hence is regarded as the principle aroma compound contributing to the aroma character of rice. In addition to aroma, cooking and eating qualities of rice are valuable properties, especially in Asia, where it is consumed as staple food.

This book is the original research work that provides deep insights on the diversity, grain morphology, physicochemical and cooking quality assessment, aroma volatile profiles, genetic diversity assessment, and marker validation for important quality parameters with special reference to the aromatic rice cultivars cultivated in the state of Maharashtra and part of Karnataka. Through personal collections and rice research station collections, we documented a total of 124 scented rice cultivars. For the first time our study listed seven landraces – Girga, Kothmirsal, Kala bhat, Chimansal, Jiri, Kalsal and Velchi from Maharashtra, and Kali kumud from Belgaum – indigenous to these regions. The grain morphological characteristics of 88 cultivars have been carried out, and the representative cultivars belonging to basmati (3) and non-basmati (57) groups were characterized for their physicochemical properties and cooking characteristics.

For qualitative and quantitative analysis of volatile compounds contributing in aroma, a highly sensitive head space-solid phase micro extraction (HS-SPME) coupled with gas chromatography flame ionization detector (GC-FID) technique was used, and a standard method was developed. Using this method, 35 marketed rice samples and 91 scented cultivars from the core collection were analyzed. Quantitative analysis of 23 major volatiles was done in both categories. In marketed samples, rice types (Basmati, Ambemohar, Kolam, Indrayani, and local) significantly contributed to the variation in 2AP, hexanal, nonanal, decanal, benzyl alcohol, vanillin, guaiacol, and indole. The method developed suffices the need of rapid and reliable method for quantification of 2AP and other aroma-related volatiles from aromatic rice cultivars. Among the 91 cultivars, 2AP, octanal, nonanal, decanal, benzyl alcohol, 2-phenylethanol, 2 amino acetophenone, indole, 1-hexanol, 1-octen-3-ol, and nonanoic acid exhibited significant variation in concentration. The study revealed that non-basmati scented landraces can endow better prospects in aromatic rice research owing to diversity in volatile composition and adaption of local conditions.

Genetic diversity assessment among the cultivars using Amplified Fragment Length Polymorphism (AFLP) revealed a high level of genetic diversity among the non-basmati groups. Validation of markers associated with selected grain quality traits in scented rice cultivars was done with respect to aroma, apparent amylose content, and grain length. The *badh2* marker for aroma allowed satisfactory discrimination of basmati from non-scented cultivars and identification of heterozygotes with the exception of 16 non-basmati scented cultivars. The SSR marker linked with *Wx* for apparent amylose content could segregate scented rice cultivars. RM-431 is linked with QTL for grain length, and no correlation was shown between allele size of RM-431 amplified and grain length.

We hope that this book will be useful to the rice scientists and breeders working on the improvement of aromatic cultivars. It will also be a guideline for qualitative and quantitative assessment of aroma volatiles involved in pleasant rice aroma.

Pune, Maharashtra, India — Altafhusain Nadaf
Sarika Mathure
Mumbai, Maharashtra, India — Narendra Jawali

Acknowledgments

At the outset, we heartily express our gratitude to the most beneficent and most merciful Almighty for giving us inspiration and capacity to explore in His fragrant creation. We heartily thank our Honorable Vice-Chancellor Professor W. N. Gade, Savitribai Phule Pune University, Pune, for his constant encouragements and support. This work has been carried out through the financial support from various agencies like the Department of Science & Technology, New Delhi, under fast track grant for young investigators to one of the authors (ABN), Savitribai Phule Pune University-Bhabha Atomic Research Centre (SPPU-BARC) collaborative research program, and the Board of College and University Development (BCUD), Savitribai Phule Pune University, Pune. We are thankful to Prof. Sujata. S. Bhargava, Head and former heads of the Department of Botany, Savitribai Phule Pune University, Pune, and head of the MBD Division, BARC, Mumbai, for providing us necessary facilities.

We express our gratitude to the agricultural rice research station at Radhanagari, Kolhapur; rice research station at Shindevahi, Chandrapur; rice research station at Karjat, Raigadh; agricultural research station at Sirsi, Karwar; rice research station at Mugad, Dharwad; and Indian Agricultural Research Institute, New Delhi, for providing rice accessions for research work. We would like to thank the assistant director of research at the Karjat Rice Research Station, Karjat, for providing field facilities. Thanks are also due to Grahak Peth supermarket for providing marketed rice samples. We would like to thank Dr. P. Srinivas, CFTRI, Mysore, for providing the standard of 2 acetyl-1-pyrroline.

We are indebted to Prof. R. J. Thengane, Advisor, Green Gold Seeds Ltd., Aurangabad, and Dr. Ajay Saini, Scientist, Molecular Biology Division, Bhabha Atomic Research Centre, Trombay, Mumbai, for their valuable suggestions and support.

We thank Prof. Robert Henry, Director, Queensland Alliance for Agriculture and Food Innovation (QAAFI), University of Queensland, Australia for writing foreword for our book.

We are thankful to the team of Springer India (Pvt.) Ltd. Life Sciences, New Delhi, for publishing this book. Finally we thank our respective family members for their patience, support, and blessings.

Authors

Contents

About the Authors

Altafhusain Nadaf is working as an associate professor at the Department of Botany, Savitribai Phule Pune University, Pune, in the area of biochemistry and molecular genetics of scented rice for the past 13 years. He was awarded Erasmus Mundus Action 2 India4EU II scholarship to visit the University of Bologna, Italy, as a visiting professor for one month (2014) and DST-BOYSCAST fellowship to work as a visiting fellow at the Centre for Plant Conservation Genetics, Southern Cross University, Lismore, Australia, for one year in 2010–2011. He has received research grants by several funding agencies including DST Fast Track Scheme for Young Scientists and DBT Rapid Grant for Young Investigators (RGYI). He has successfully guided several Ph.D. and M.Phil. students and women scientists (under DST Women Scientists Scheme – A and B). He is working as a reviewer for many international journals. He has in his credit more than 45 research papers published in peer-reviewed national and international journals of repute. He has presented his research work at national and international level.

Sarika Mathure is working on characterization of Indian scented rice cultivars with respect to their quality traits, diversity, and molecular markers. She has obtained her doctoral degree in biotechnology from Savitribai Phule Pune University, Pune. She has published 6 research papers in international refereed journals. She is involved in quantitative analysis of rice volatiles. Her work was presented in several international and national conferences.

Narendra Jawali former head of the Molecular Biology Division, Bhabha Atomic Research Centre, Mumbai, India, obtained his doctorate degree in biochemistry from Mumbai University. He was member secretary of Basic Sciences Committee, Board of Research in Nuclear Sciences, a research funding agency of the Government of India. He carried out research in the area of biochemistry and molecular genetics of plants. His investigations on rRNA sequences have increased understanding of molecular phylogeny and evolution of species belonging to subgenera *Vigna* and *Ceratotropis* of the genus *Vigna*. His studies in wheat have led to a genetic linkage map, QTL map, and SCAR markers for a rust resistance gene. His collaborative studies have resulted in understanding the mechanism of salt tolerance in rice and revealed quality and genetic diversity among some non-basmati scented rice from south India. His studies have revealed the mechanism of electron transport and role of Cytb557 in nitrate reductase and the reaction mechanism and regulation of phosphoenolpyruvate carboxylase and malic enzyme from maize.

1 Introduction

Rice is the staple food of around three billion people, most of them in Asia, which accounts for 90 % of global rice consumption. Rice constitutes a major source of nutrition and contributes a significant share of dietary energy in a number of Asian countries. Among 23 species of genus *Oryza*, *Oryza sativa* L. is cultivated in Asia and *O. glaberrima* Steud. in West Africa (Vaughan et al. 2003). *O. sativa* L. is further differentiated into *indica* and *japonica* (Khush 2000). Rice has immense diversity, and it is estimated that more than 100,000 varieties of rice exist in the world (Khush 1997). India has an ancient heritage of rice cultivation and has over 70,000 cultivars of rice germplasm (Siddiq 1992).

1.1 Scented Rice

For many centuries, aromatic rice has been preferred for consumption because of its pleasant aroma and unique taste. The tropical countries have their own set of scented rice varieties that are popularly consumed, e.g., Basmati, Jasmine, KDML, Della, etc. Therefore, the rice world market ranks aromatic rice at the top, which is sold at a premium price in local and export markets (Sakthivel et al. 2009). Scented rice constitutes a small but a special group of rice, considered as of best quality. The Indian subcontinent flourishes with hundreds of indigenous aromatic cultivars and landraces, and the diversity of scented rice of India is highest in the world (Singh and Singh 2003). Scented rice is further classified as basmati and nonbasmati types. Basmati rice – the scented pearl – is nature's gift exclusively to the Indian subcontinent. Originating in the foothills of the Himalayas, Basmati rice is characterized by extra long slender grain, pleasant and distinct aroma, and soft and fluffy texture of the cooked rice (Nagaraju et al. 2002). Basmati type is long slender grain; it exhibits kernel length above 6.6 mm, L/B ratio of more than 3, and high kernel elongation after cooking (ratio above 1.8). These unique features of basmati are said to be due to the culmination of centuries of selection and cultivation by farmers, which are well preserved and maintained in their purest form as traditional basmati varieties (Siddiq et al. 2012). In addition to basmati varieties, many indigenous nonbasmati scented rice varieties are also locally cultivated. In a compendium published by Singh and Singh (2003), the authors mentioned that the diversity of scented rice from India is the largest in the world. Majority of the indigenous scented rice cultivars are small and medium grained (Singh et al. 2000). Though the nonbasmati types do not possess all the characteristics of basmati, these varieties excel equally as far as aroma and other characters are concerned. Hence, nonbasmati varieties are traded popularly in the market, e.g., Ambemohar, Ghansal in Maharashtra, Kalanamak, Bindli in Uttar Pradesh, etc. However, unfortunately, these have somehow not

A. Nadaf et al., *Scented rice (Oryza sativa L.) Cultivars of India: A Perspective on Quality and Diversity*, DOI 10.1007/978-81-322-2665-9_1

got the attention of rice scientists and traders, including exporters, to the extent that basmati has. The introduction of hybrid rice varieties has further contributed in reduction in area under cultivation of nonbasmati scented rice. As a result, most of this valuable wealth has either already vanished or is on a decline. The cultivation of nonbasmati scented rice is now confined to limited pockets where farmers grow them either for self-consumption or for special occasions. Since local varieties are a rich source of gene pool (Malik et al. 1994), it is vital to collect, characterize, document, and conserve these varieties.

1.2 Scented Rice Quality Parameters

Rice quality, in general, is a complex trait comprising many physicochemical characteristics. As rice is consumed as whole grain, quality parameters such as appearance in terms of grain shape and size, along with its cooking and eating qualities, play a critical role in consumer preference. Raychaudhari (1992) stated that in ancient India, rice was distinguished into three kinds according to hardness, color, flavor, and size of grain. Preferences for grain size and shape vary with consumers as some ethnic groups prefer short bold grains, while medium and long slender grains are preferred by others. In general, rice with soft to medium gel consistency, intermediate amylose content, and intermediate gelatinization temperature is a preferred level for consumers (Khush et al. 1979). However, the cooking quality preferences are known to vary within the country, within the ethnic groups, from one country to another, and within different geographical regions (Juliano et al. 1964). Nonbasmati rice usually has medium to short grains and excels in the demand of local market than long grain rice. Some of the outstanding examples of indigenous nonbasmati scented varieties are Kalanamak (popularly called the "black pearl of Utter Pradesh"), Shakarchini, and Hansaraj of Utter Pradesh; Dubraj and Chinoor of Chhattisgarh; Kalajoha of northeast; Ambemohar of Maharashtra; and Randhunipagal of Orissa and West Bengal region. Bindli, a small-grained scented variety of Uttar Pradesh, is known to possess aroma and cooking qualities much superior to basmati. It elongates 200 % on cooking, surpassing basmati in this respect. Hence, characterization of cultivars with respect to their quality traits will help for the possible use of these varieties for commercialization and in breeding program for the development of quality rice.

1.3 Rice Aroma

The rice aroma constitutes more than 100 volatile compounds belonging to hydrocarbon (13), acid (14), alcohol (13), aldehyde (16), ketone (14), ester (8), and phenol (5) categories (Hussain et al. 1987; Bergman et al. 2000). Among these volatile compounds, 2-acetyl-1-pyrroline (2AP), a-pyrrolidone, pyridine, guaiacol, indole, *p*-xylene, and 1-octen-3-ol are favoring and enhancing the consumer acceptability of rice, while lipid oxidation products, such as hexanal, acetic acid, and pentanoic acid laid negative influence on consumer acceptability. The interactive effect of these volatile compounds collectively influences the aroma character among different scented rice varieties (Yang et al. 2008a; Maraval et al. 2008). Thirteen odor-active compounds, 2AP, hexanal, (*E*)-2-nonenal, octanal, heptanal, nonanal, 1-octen-3-ol, (*E*)- 2-octenal, (*E*,*E*)-2,4-nonadienal, 2-heptanone, (*E*,*E*)-2,4-decadienal, decanal, and guaiacol, have been identified as primary compounds responsible for inducing variations in the aroma in six scented rice varieties (Yang et al. 2008b). The volatile aroma compounds in general and 2AP in particular are synthesized in aerial parts of rice seedlings (Yoshihashi 2002).

1.3.1 2-Acetyl-1-Pyrroline as a Principle Aroma Compound

The chemical nature of the aroma compound remained a mystery until year 1982, when Ron Buttery with his coworkers successfully elucidated the fragrance and identified 2-acetyl-1-pyrroline (2AP) as the principal aroma compound

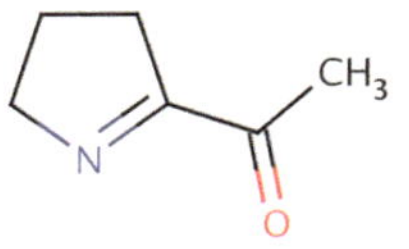

Fig. 1.1 Structure of 2-acetyl-1-pyrroline

(Fig. 1.1) (2AP; IUPAC names 5-acetyl-3,4-dihydro-2H-pyrrole and 1-(3,4-dihydro-2H-pyrrol-5-yl)-ethanone) responsible for the pleasant aroma (Buttery et al. 1982). The universality of this molecule was confirmed by subsequent rice aroma analyses in several aromatic varieties worldwide. The importance of 2AP was realized when presence of 2AP was detected beyond scented rice varieties from unicellular bacteria to urine of tiger. Since its discovery, 2AP has not ceased to reveal its presence in a vast variety of food products and also in wide range of living systems (Adams and Kimpe 2006). Interestingly, besides living systems, in a large number of heated and processed food products, 2AP has been found as an important Maillard flavor compound (Adams and Kimpe 2006; Demyttenaere et al. 2002). Buttery et al. (1983b) described the odor of 2AP as "popcorn" like aroma with a threshold of 0.1 part of compound per 109 parts of water. Later they also identified and reported 2AP in *P. amaryllifolius* leaves in order of 1 ppm, 10 fold more than that found in milled scented rice varieties and 100 fold than that found in common rice (Buttery et al. 1983a). In the subsequent years, presence of 2AP as a principal aroma compound was confirmed in basmati, jasmine, and other scented rice varieties (Buttery et al. 1982, 1986, 1988; Tanchotikul and Hsieh 1991; Lin et al. 1990; Paule and Powers 1989; Widjaja et al. 1996).

The pyrroline ring in 2AP makes the molecule highly unstable (Buttery et al. 1982). Buttery et al. (1982) succeeded in preserving synthetic 2AP in acidic form as long as 3 months. Its synthesis required costly catalysts like rhodium on alumina with very low final yield. Efforts by Brahmachary (1996) showed that 2AP could be preserved for more than 6 months as 2AP citrate salt by treating with citric acid. De Kimpe and Keppens (1996) synthesized diethylacetal of 2AP, which was claimed to be more stable. Srinivas et al. (2004) successfully stabilized this compound by making the derivative of 2AP using acids. The compound was found to be stable for more than 1 year (Srinivas et al. (2004)). The unstable nature of compound remained as a main constrain for its commercial synthesis and its use in flavor industry.

1.3.2 Genetic and Molecular Basis of 2AP Biosynthesis

Several attempts have been made to reveal the genetic and molecular basis of 2AP formation in aromatic rice. Fragrance in rice is a highly heritable trait. The genes controlling aroma trait in rice have been studied by genetic and molecular mapping. Siddiq et al. (1986) studied an Indian scented rice variety and found two recessive genes for aroma, located on chromosome 5 and 9. Subsequently, this was also confirmed by Tsuzuki and Shimokawa (1990). Then numbers of genes controlling the inheritance of aromatic characters were reported to vary from one to four pairs (Lin et al. 1990; Pinson 1994). Pinson (1994) determined the genes controlling aroma in six cultivars and found that the scent of aromatic rice was under the control of one or two recessive genes, dependent upon the cultivar studied. However, among some rice varieties, aroma was found to be controlled by a single recessive gene (Berner and Hoff 1986; Yano et al. 1991). Further studies identified a single recessive gene located on chromosome 8 by using different techniques, viz., restriction fragment length polymorphism (RFLP) (Ahn et al. 1992; Garland et al. 2000), translocation and trisomic lines from nonfragrant rice cultivar IR36 (Li et al. 1996), molecular markers RG28 and RG1 (Lorieux et al. 1996), and SSR markers RM210 and RM515 (Ren et al. 2004). Finally, physical fine mapping study and the availability of rice genome sequence revealed that a candidate gene (*fgr/badh2/Os2AP*) homologous to betaine aldehyde dehydrogenase (*badh*) on chromosome 8 is responsible for aroma formation in aromatic rice (Cordeiro et al. 2002; Wanchana et al. 2005; Bradbury et al. 2005).

An eight-base pair deletion in exon 2 or 7 of *badh2* gene on chromosome 8 results in a loss of function of the encoded enzyme *badh2* (Wanchana et al. 2005; Bradbury et al. 2005; Shi et al. 2008; Fitzgerald et al. 2008), and consequently 2AP accumulates in fragrant rice. This was recently confirmed by the absence of 2AP in an aromatic variety after transformation with functional *badh* (Chen et al. 2008), and suppression of *badh2* transcript in a nonaromatic rice increases 2AP concentration (Niu et al. 2008). Based on this recessive gene (*fgr*)/*badh2*, PCR-based codominant functional markers for fragrance trait were developed and exploited for the identification of fragrant and nonfragrant rice genotypes (Sakthivel et al. 2009). However, deletion in the *badh2* gene is not the only cause of aroma in all aroma rice varieties; aromatic rice of different origins may have various regulation or at least one other mutation that drives the accumulation of 2AP (Fitzgerald et al. 2008).

1.4 Utility of Molecular Markers in Determination of Quality Parameters

Use of molecular markers is considered best for the analysis of genetic diversity and varietal identification, since there is no effect of stage of development and environmental and management practices. Restriction fragment length polymorphism (RFLP) and polymerase chain reaction (PCR)-based techniques like random amplified polymorphic DNA (RAPD), simple sequence repeat (SSR), and amplified fragment length polymorphism (AFLP) have been utilized in assessing diversity in *O. sativa* L. In the Indian subcontinent, studies on genetic diversity analysis are focused mainly on the basmati genotypes. Comparatively, nonbasmati indigenous scented rice genotypes have received little attention and only few studies were undertaken with respect to their genetic diversity (Prashanth et al. 2002, Jain et al. 2004; Joshi and Behera 2006). Also, detection of unique markers specific to genotypes could serve as marker tag for these scented rice varieties. Therefore, to address the need of diversity analysis of rice cultivars grown in Maharashtra and Karnataka and their comparison with other nonbasmati scented rice, nonscented rice, and basmati rice cultivars from India, the present study was undertaken.

New developments in genomics have provided tools for discovering and tagging novel alleles and offered great promise to understand the inheritance of grain quality traits and crop improvement through their use in marker-assisted selection (MAS) (Xu and Crouch 2008). In rice, several molecular markers have been developed and found as suitable in MAS (Jena and Mackill 2008). These markers need to be validated by testing the reliability of the marker in determining the target phenotype in independent populations and different genetic backgrounds.

In view of these, the present study was undertaken with a major focus on local nonbasmati scented rice cultivars of Maharashtra and Karnataka with respect to grain quality determination, qualitative and quantitative analysis of aroma volatiles, diversity analysis using molecular markers, and validation of markers associated with grain quality.

References

Adams A, Kimpe ND (2006) Chemistry of 2-Acetyl-1-Pyrroline, 6-Acetyl-1,2,3,4-Tetrahydropyridine, 2-Acetyl-2-Thiazoline and 5-Acetyl-2,3-Dihydro-4H-thiazine: extraordinary Maillard flavor compounds. Chem Rev 106:2299–2319

Ahn SN, Bollisch CN, Tanksley SD (1992) RFLP tagging of a gene for aroma in rice. Theor Appl Genet 84:825–828

Bergman C, Delgado J, Bryant R, Grimm C, Cadwallder K, Webb BA (2000) Rapid gas chromatographic technique for quantifying 2-Acetyl-1-Pyrroline and hexanal in rice *Oryza sativa* L. Cereal Chem 77(4):454–458

Berner DK, Hoff BJ (1986) Inheritance of scent in American long grain rice. Crop Sci 26:876–878

Bradbury LM, Fitgerald TL, Henry RJ, Jin Q, Waters DLE (2005) The gene for fragrance in rice. Plant Biotechnol J 3:363–370

Brahmachary RL (1996) The expanding world of 2-acetyl-1-pyrroline. Curr Sci 71(4):257–258

Buttery RG, Ling LC, Juliano BO (1982) 2-acetyl-1-pyrroline: an important aroma component of cooked rice. Chem Ind 23:958

Buttery RG, Juliano BO, Ling LC (1983a) Identification of rice aroma compound 2-acetyl-1-pyrroline in Pandan leaves. Chem Ind 23:478

Buttery RG, Ling LC, Juliano BO, Turnbaugh JG (1983b) Cooked rice aroma and 2-acetyl–1-pyroline in rice. J Agric Food Chem 31:823–826

Buttery RG, Ling LC, Mon TR (1986) Quantitative analysis of 2-acetyl-1-pyrroline in rice. J Agric Food Chem 34:112–114

Buttery R, Turnbaugh J, Ling L (1988) Contributions of volatiles to rice aroma. J Agric Food Chem 36:1006–1009

Chen S, Yang Y, Shi W, Ji Q, He F, Zhang Z, Xu M (2008) Badh2, encoding betaine aldehyde dehydrogenase, inhibits the biosynthesis of 2-acetyl-1-pyrroline, a major component in rice fragrance. Plant Cell 20(7):1850–1861

Cordeiro GM, Christopher MJ, Henry RJ, Reineke RF (2002) Identification of microsatellite markers for fragrance in rice by analysis of the rice genome sequence. Mol Breed 9:245–250

De Kimpe N, Keppens M (1996) Novel syntheses of the major flavour components of bread and cooked rice. J Agric Food Chem 44(6):1515–1519

Demyttenaere J, Abbaspour Tehrani K, De Kimpe N (2002) The chemistry of the most important flavour compounds of bread and cooked rice. ACS Symp Ser 826:158–173

Fitzgerald MA, Sackville Hamilton NR, Calingacion MN, Verhoeven HA, Butardo VM (2008) Is there a second fragrance gene in rice? Plant Biotechnol J 6(4):416–423

Garland S, Lewin L, Blakeney A, Reineke R, Henry R (2000) PCR based molecular markers for the fragrance gene in rice (*Oryza sativa* L.). Theor Appl Genet 101:364–371

Hussain AA, Maurya DM, Vaish CP (1987) Studies on quality status of indigenous upland rice (*Oryza sativa*). Indian J Genet 47(2):145–152

Jain S, Jain RK, McCouch SR (2004) Genetic analysis of Indian aromatic and quality rice (*Oryza sativa* L.) germplasm using panels of fluorescently-labeled microsatellite markers. Theor Appl Genet 109:965–977

Jena KK, Mackill DJ (2008) Molecular markers and their use in marker-assisted selection in rice. Crop Sci 48(4):1266–1276

Joshi RK, Behera L (2006) Identification and differentiation of indigenous non-basmati aromatic rice genotypes of India using microsatellite markers. Afr J Biotechnol 6(4):348–354

Juliano BO, Bautista GM, Lugay JC, Reyes AC (1964) Studies on the physicochemical properties of rice. J Agric Food Chem 12:131–138

Khush GS (1997) Origin, dispersal, cultivation and variation of rice. Plant Mol Biol 35(1):25–34

Khush GS (2000) Taxonomy and origin of rice. In: Singh RK, Singh US, Khush GS (eds) Aromatic rices. Oxford & IBH Pub. Co., New Delhi, pp 5–13

Khush GS, Paule CM, Delacruz NM (1979) Rice grain quality evaluation and improvement at IRRI. In: Chemical aspects of rice grain quality. IRRI, Los Baños, pp 21–31

Li J, Ku DF, Li LF (1996) Analysis of fragrance inheritance in scented rice variety Shenxiangjing-4. Acta Agric Shanghai 12:78–81

Lin H (1995) RFLP mapping of QTLs for grain shape traits in indica rice (*Oryza sativa* L. subsp. *indica*). Sci Agric Sin 28:1–7

Lin C, Hsih T, Hoff B (1990) Identification and quantification of the "Popcorn"-like Aroma in Louisiana Aromatic Delia Rice (Oryza sativa, L.). J Food Sci 55:1466–1467

Lorieux M, Petrov M, Huang N, Guiderdoni E, Ghesquière A (1996) Aroma in rice: genetic analysis of a quantitative trait. Theor Appl Genet 93:1145–1151

Malik SS, Dikshit N, Dash AB, Lodh SB (1994) Studies on agromorphological and physicochemical characteristics of local scented rice varieties. Oryza 31:106–110

Maraval I, Mestres C, Pernin K, Ribeyre F, Boulanger R, Guichard E, Gunata Z (2008) Odor-active compounds in cooked rice cultivars from Camargue (France) analyzed by GC#O and GC#MS. J Agric Food Chem 56(13):5291–5298

Nagaraju J, Kathirvel M, Kumar RR, Siddiq EA, Hasnain SE (2002) Genetic analysis of traditional and evolved Basmati and non-Basmati rice varieties by using fluorescence-based ISSR-PCR and SSR markers. Proc Natl Acad Sci U S A 99(9):5836–5841

Niu X, Tang W, Huang W, Ren G, Wang Q, Luo D, Xiao Y, Yang S, Wang F, Lu BR, Gao F, Lu T, Liu Y (2008) RNAi-directed downregulation of Osbadh2 results in aroma (2-acetyl-1-pyrroline) production in rice (*Oryza sativa* L.). BMC Plant Biol 8:100

Paule CM, Powers JJ (1989) Sensory and chemical examination of aromatic and non aromatic rices. J Food Sci 54:343–346

Pinson SRM (1994) Inheritance of aroma in six rice cultivars. Crop Sci 34:1151–1157

Prashanth SR, Parani M, Mohanty BP, Talame V, Tuberosa R, Parida A (2002) Genetic diversity in cultivars and landraces of Oryza sativa subsp. indica as revealed by AFLP markers. Genome 45(3):451–459

Raychaudhari P (1992) Agriculture in ancient India. ICAR, New Delhi, p 62

Ren JS, Xiao PC, Chen Y, Huang X, Wu XJ, Wang XD (2004) Study on heredity of aroma genes in several maintainer lines of aromatic rice. Seed 23:24–28

Sakthivel K, Sundaram RM, Shobharani N, Balachandran SM, Neeraja CN (2009) Genetic and molecular basis of fragrance in rice. Biotechnol Adv 27(4):468–473

Shi W, Yang Y, Chen S, Xu M (2008) Discovery of a new fragrance allele and the development of functional markers for the breeding of fragrant rice varieties. Mol Breed 22(2):185–192

Siddiq EA (1992) Rice in India – potential areas for research and development. Oryza 29:1–6

Siddiq EA, Sadananda AR, Zaman FU (1986) Use of primary trisomic of rice in genetic analysis. In: Genetics

R (ed) Proceedings of the international rice genetics symposium. International Rice Research Institute, Manila, pp 185–197

Siddiq EA, Vemireddy LR, Nagaraju J (2012) Basmati rices: genetics, breeding and trade. Agric Res 1(1):25–36

Singh RK, Singh US (2003) A treatise on the scented rices of India. Kalyani Publishers, New Delhi

Singh RK, Singh US, Khush GS, Rohilla R, Singh JP, Singh G, Shekhar KS (2000) Small and medium grained scented rices of India. In: Singh RK, Singh US, Khush GS (eds) Aromatic rices. Oxford & IBH Pub. Co, New Delhi, pp 155–177

Srinivas P, Sulochanamma G, Raghavan B, Gurudutt K (2004) An improved process for the stabilization of 2-acetyl-1-pyrroline, the basmati rice flavourant. WIPO Patent, WO/2004/056202

Tanchotikul U, Hsieh TC-Y (1991) An improved method for quantification of 2-acetyl-1-pyrroline, a "popcorn"- like aroma, in aromatic rice by high-resolution gas chromatography/mass spectrometry/selected ion monitoring. J Agric Food Chem 39:944–947

Tsuzuki E, Shimokawa E (1990) Inheritance of aroma in rice. Euphytica 46:157–159

Vanavichit A (2007) Molecular diversity and evolution of aromatic rice in Thailand. In: International training workshop, the conservation and utilization of tropical/subtropical plant genetic resources. Taiwan Agricultural Research Institute, Council of Agriculture (TARIC), Taiwan, pp 131–134

Vaughan DA, Morishima H, Kadowaki K (2003) Diversity in the Oryza genus. Curr Opin Plant Biol 6(2):139–146

Wanchana S, Kamolsukyunyong W, Ruengphayak S, Toojinda T, Tragoonrung S, Vanavichit A (2005) A rapid construction of a physical contig across a 4.5 cM region for rice grain aroma facilitates marker enrichment for positional cloning. Sci Asia 31:299–306

Widjaja RW, Craske JD, Wootton M (1996) Comparative studies on volatile components of non-fragrant and fragrant rices. J Sci Food Agric 70:151–161

Xu Y, Crouch JH (2008) Marker-assisted selection in plant breeding: from publications to practice. Crop Sci 48:391–407

Yang DK, Lee KS, Jeong OY, Kim KJ, Kays SJ (2008a) Characterization of volatile aroma compounds in cooked black rice. J Agric Food Chem 56(1):235–240

Yang DS, Robert LS, Kyu-Seong L, Stanley JK (2008b) Comparison of odor-active compounds from six distinctly different rice flavor types. J Agric Food Chem 56(8):2780–2787

Yano M, Shimosaka E, Sato A, Nakagahra M (1991) Linkage analysis of a gene for scent in *indica* rice variety, surjamkhi, using restriction fragment length polymorphism makers (in Japanese). Jap J Breed 41:338–339

Yoshihashi T (2002) Quantitative analysis on 2-Acetyl-1-pyrroline of an aromatic rice by stable isotope dilution method and model studies on its formation during cooking. J Food Sci 67(2):619–622

2 Collection of Germplasm of Scented Rice Cultivars and Grain Morphological Assessment

2.1 Introduction

Ancient India is one of the oldest regions where cultivation of *O. sativa* L. began. The foothill of the Himalayas is the center of diversity of scented rice of Group V; from here by westward route scented rice cultivars are distributed in Rajasthan, Madhya Pradesh, Maharashtra, and Gujarat (Glaszmann 1987). In these states, numerous scented varieties belonging to this group are grown under different names (Khush 2000, Table 2.1). It is estimated that India has over 70,000 cultivars of rice germplasm and has a sizable number of wild forms still to be collected and conserved (Siddiq 1992). Since the time of civilization, thousands of locally adapted scented rice genotypes have evolved as a consequence of natural and human selection. These landraces are the genetic reservoirs of useful genes. The collection and evaluation of landraces are a part of the fundamental work of rice geneticists for breeding purposes. Considering the need for broadening the gene pool of rice, it is necessary to collect and conserve the cultivars that are locally cultivated and maintained by farmers.

In view of collecting the germplasm, a special drive that was undertaken for upland paddy varieties cultivated in Maharashtra, Madhya Pradesh, and five other states resulted in the collection of 1938 cultivars. The National Bureau of Plant Genetic Resources (NBPGR), New Delhi, India, further augmented the collections during 1983–1989 by about 4862 cultivars. Parts of Maharashtra were also explored during joint explorations led by Kihara in the early 1960s and Watabe in the late 1960s and early 1970s (Singh et al. 2000b). Even though all these explorations have resulted in a large assembly of scented rice cultivars maintained as working/active collections/base collections at various gene banks, majority of the cultivars lack precise information on important traits (Rana et al. 2009).

The large-scale spread of modern, high-yielding varieties has replaced the traditional varieties, especially in the irrigated rice ecosystem leading to reduced genetic base and thus increased genetic vulnerability. In the past few decades, increase in the share of high-yielding varieties and shrinkage in the area of local varieties have been reported in India (Hore 2005, Patil and Patil 1994, Rana et al. 2009), as well as in several other countries (Bhatti 1998, Chaudhary et al. 2006, Itani 1993, Juma 1989).

Collection and characterization of cultivars are not only important for utilizing the appropriate attribute based donors in breeding programs but also essential in the present era for protecting the unique rice. Thus, considering this background and the need for conservation, efforts were undertaken to collect rice cultivars, especially nonbasmati scented cultivars from Maharashtra and parts of Karnataka.

A. Nadaf et al., *Scented rice (Oryza sativa L.) Cultivars of India: A Perspective on Quality and Diversity*, DOI 10.1007/978-81-322-2665-9_2

Table 2.1 State-wise aromatic rice (*Oryza sativa* L.) cultivars of India (part of the information taken from Singh and Singh (2003))

State	Cultivar
Basmati types	
Jammu & Kashmir	Basmati 370
Himachal Pradesh	Desi basmati – 23
Haryana	Tarori basmati, Karnal local, Basmati 370
Punjab	Basmati 370, Basmati 217, Muskan, Begami, Basmati 6129
Uttar Pradesh	Basmati 370, Katarani, Dehradun basmati, Hansraj
Rajasthan	Basmati Kota, Basmati local, Kali Kamod, Nawabi Kolam, Nayan Kajal
Madhya Pradesh	Madhu Malti
Nonbasmati types	
Jammu & Kashmir	Mushk Budgi
Assam	Badashahbhog, Prasadbhog, Malbhog, Bhabeli Joha, Kala Joha, Manaki Joha, Bagri Bhog, Kali Gira Bhog
Bihar	Badashahbhog, Ram Tulsi, Mohanbhog, Tulsimanjari, NP 49
Uttar Pradesh	Kalanamak, Tilakchandan, Bindli, Vishnuparag, Shakarchini, Lalmati, Badshahpasand, Laung Choor, Ramjawain, Bahantaphul, Tulsiprasad
Gujarat	Kamod 118, Pankholi 203, Zeerasal
Madhya Pradesh	Dubraj 354, Chinnor, Kali Kamod, Chattri, Baspatri, Chinni Sagar, Chinni Gauri, Jauphool, Adamchini, Amarjyoti, Badashahbhog (R 14), Bhataphool, Kalimuch, Shakarchini, Vishnu Bhog
Maharashtra	Kala bhat, Kate chinoor, Kondhekar chinoor, Parabhani chinoor, Makarand, Ambemohar Ajra, Champakali, Ghansal, Girga, Kothmirsal, Chimansal, Kamod, Indrayani, Ambemohar, Ambemohar Pandhara, Ambemohar-Tambda, Jiri, Kalsal, Kamavatya, Khadkya, Lal bhat, Lal dodki, Raibhog, Tamsal, Ambemohar-157, Gham, Velchi, Pusa sugandha

(continued)

Table 2.1 (continued)

State	Cultivar
Karnataka	Luchai, Mothi luchai, Navedhan, Suvarna wild
	Ratnagiri 24, Sahyadri, Sonam, Sonamasuri
	Dev bhat, Gari kolpi, Hali kolpi, Kolamb, Majeri, Pandhara dodki, Mahsuri, Zeeni kolam
	Swarna, Chitak bhat, Manila
Orissa	T 812, Badshabhog, T412, Barangomati, Karpurkali, Sagartara, Takurbhoga
Andhra Pradesh	Amritsari (HR 22), Sukhdas (HR 47), Kaki Rekhuli (HR 59)
Tamilnadu	Geerga Samba
West Bengal	Gopalbhog, Sitabhog, Govindabhog, Kamininbhog, Tulaipanja
Tripura	Govindbhog (White), Govindbhog (Black), Sada Khasa, Kala Khasi, Kali Jira, Koliajiri
Mizoram	Tai, Farate, Bawng Buh, Mawng Buh, Zongam, Phanrai

2.2 Methodology and Approaches

2.2.1 Exploration and Collection of Rice Cultivars

The collection of rice cultivars was made by field collection, procurement of cultivars from rice research stations and research institutes, and purchase of cultivars.

2.2.1.1 Field Collection

Rice-growing regions of Maharashtra state and Belgaum district, Karnataka, were surveyed in years 2005–2009 for scented rice cultivars. During this period, local farmers were enquired primarily about native scented landraces. Based on this information, fine grid survey was conducted in Konkan belt (Thane, Raigad, Ratnagiri, and Sindhudurg districts), West Maharashtra (Nashik, Ahamadnagar, Pune, and Kolhapur), and Vidharba (Bhandara, Chandrapur, Gudchiroli,

Gondia districts) for collection of cultivars. Fresh seed stocks for each cultivar were collected from fields, farmer's stores, local markets, and threshing floors. Along with this, farmer's opinions about the peculiarities and shortcomings of the cultivars were also recorded. The localities of collection are indicated in Fig. 2.1.

2.2.1.2 Procurement from Research Stations and Institutes

Germplasm procured from research stations in Maharashtra, viz., Agricultural Rice Research Station, Radhanagari, Dist. Kolhapur (ARRS, Radhanagari), and Rice Research Station, Shindevahi, Dist. Chandrapur (RRS, Shindevahi), and from research stations in Karnataka, viz., Agricultural Research Station (Paddy), Sirsi, Dist. Karwar (ARS, Sirsi), and Rice Research Station, Mugad, Dist. Dharwad (RRS, Mugad) ascertains a comprehensive collection (Fig. 2.1).

In addition, Basmati rice types were procured from the Indian Agricultural Research Institute, New Delhi, for comparative analysis. Seeds of some cultivars were also procured from National Seed Corporation Ltd., New Delhi. The cultivars were classified as scented (nonbasmati and basmati) or nonscented type and further categorized as landrace, selection, and hybrid.

2.2.2 Maintenance of Cultivars

The collected cultivars are conserved ex situ. The stocks of cultivars are maintained at −20 °C at the University of Pune, Pune, Maharashtra. Also, the cultivars are grown at Karjat rice research station field, Raigad, Maharashtra, to maintain viability.

2.2.3 Assessment of Grain Morphology

Eighty-eight scented rice cultivars, including cultivars from field collection and representatives from research stations, were assessed for grain morphology following the guidelines of the IBPGR-IRRI Rice Advisory Committee (1980). Grain morphology in terms of awning, awn color, lemma and palea characters, pubescence, color of sterile lemma, and apiculus color was recorded as described in Table 2.2.

2.3 Scented Rice Diversity Assessment in Maharashtra and Karnataka

2.3.1 Exploration and Collection of Rice Cultivars

The study resulted in the total collection of 124 rice cultivars. Among these, 52 were collected from farmer's fields, 71 were procured from research stations, and 1 was purchased.

2.3.1.1 Field Collection

Field collection consisted of 32 nonbasmati scented, 1 basmati, and 19 nonscented rice cultivars. The details of field collection are presented as Fig. 2.2 and Table 2.3.

It was observed that Western Maharashtra, particularly Pune district, possesses higher number of scented rice cultivars, followed by Kolhapur district (Fig. 2.2). Besides landraces, Ambemohar-157, a pure line selection of Ambemohar landraces, and two hybrid varieties, viz., Indrayani and Makarand, were also collected. Traditional landraces Basumati (also known as Belgaum basmati), Kali kumud, and Kumud were collected from Belgaum district, Karnataka. We, for the first time, report seven landraces – Girga, Kothmirsal, Kala bhat, Chimansal, Jiri, Kalsal, and Velchi – grown in Maharashtra and Kali kumud from Belgaum as indigenous to these regions. Ambemohar, Kamod, and Chinoor have been traditionally grown in certain pockets of Maharashtra (Nerkar et al. 2003). We could collect four cultivars of Ambemohar varying in grain morphology and quality traits, viz., Ambemohar Ajra, Ambemohar Pandhara, Ambemohar-Tambda, and Ambemohar from Pune and Kolhapur districts, and three Chinoor types, viz., Kate chinoor, Kondhekar chinoor, and Parabhani chinoor from Chandrapur district, that have not been reported earlier. Only one type of Kamod was observed to be in cultiva-

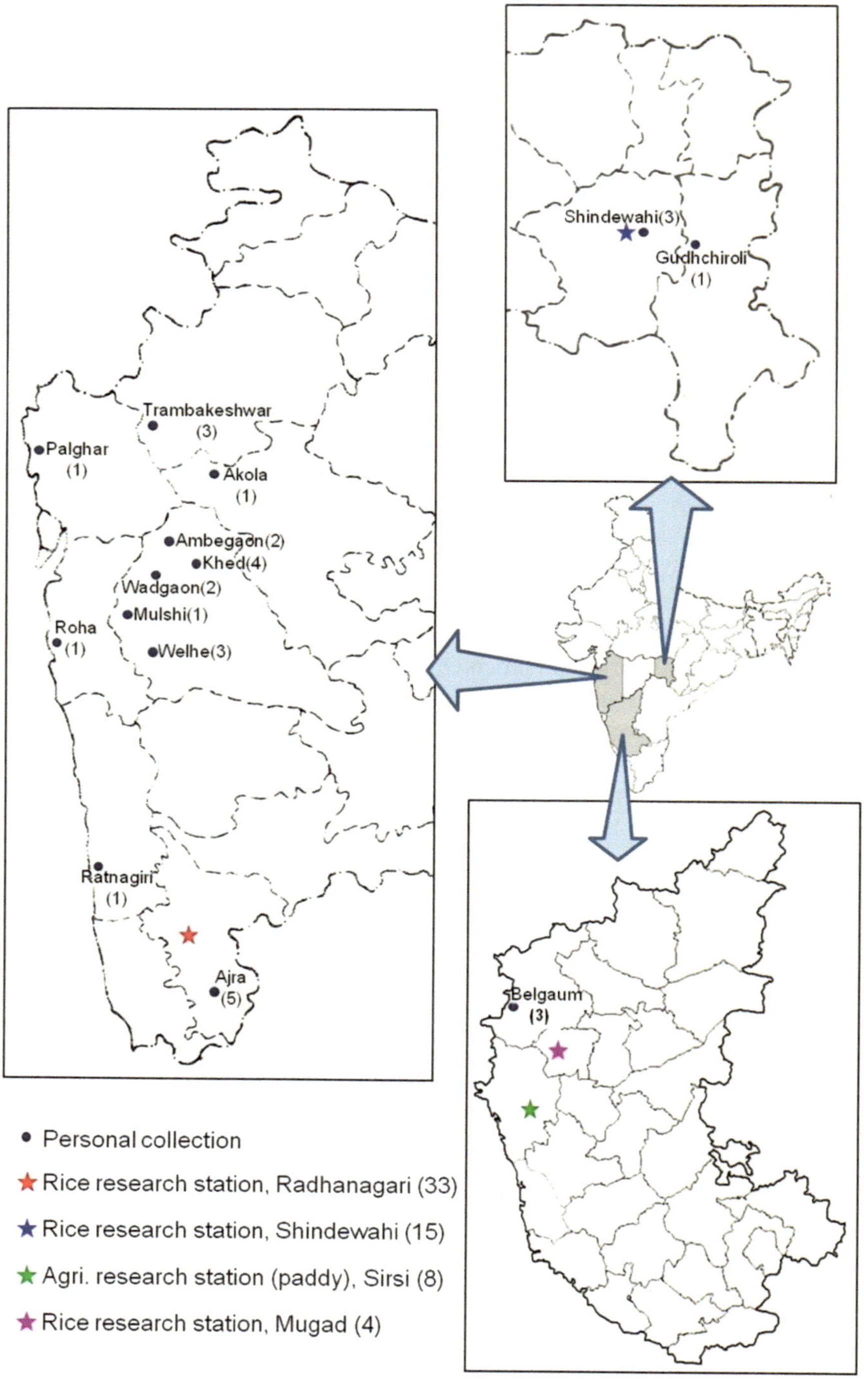

Fig. 2.1 Maps of Maharashtra and Karnataka indicating localities of collection for nonbasmati scented rice cultivars and research stations. Number of nonbasmati scented rice cultivars collected at each location is given in brackets

Table 2.2 Descriptors of rice (*O. sativa* L.) grain morphology

Character	Description with code
Awning (C1)	Absent (0), short straw colored (1), short red colored (2), short black colored (3), long straw colored (4), long red colored (5), long black colored (6)
Lemma and palea color (C2)	Straw (1), brown spots on straw (2), purple spots on straw (3), straw with furrows (4), straw with golden furrows (5), straw with brown furrows (6), straw with purple furrows (7), brown spots on straw with furrows (8), golden (9), golden with furrows (10), golden with brown furrows (11), red (12), red with furrows (13), brown (14), brown with furrows (15), black (16), black with furrows (17)
Pubescence on lemma and palea (C3)	Absent (0), present (1)
Sterile lemma color (C4)	Straw (1), gold (2), red (3), purple (4)
Apiculus color (C5)	Straw (1), red (2), purple apex (3), purple (4), brown apex (5), brown (6), black (7)

tion. Along with these, we could collect Champakali, Ghansal, Kamavatya, Khadkya, Lal bhat, Lal dodki, Raibhog, Tamsal, and Gham landraces grown in Maharashtra.

Some of the cultivars collected during study possess unique features, viz., cultivar Velchi collected from Ratnagiri, having distinct aroma resembling cardamom. Another cultivar, Gham, grows in brackish water areas, and Khadkya grows in less water. Hanamaratti et al. (2008) reported Belgaum basmati, Champakali, and Kumud from Karnataka and studied agronomy and drought resistance in Champakali. However, detailed descriptions of these cultivars were not reported by them. Improved varieties, viz., Ambemohar-157, Ambemohar-159, Ambemohar-102, Krishnasal, Banaspatri, and Chinoor, were released for cultivation in Maharashtra before 1965. Indrayani, Pawana, Phule maval, Prabhavati, Sugandha, Parag 401, PKV-HMT, and PKV-Makarand developed by hybridization were also released for cultivation in this region. However, most of these varieties have now gone out of cultivation.

Survey for scented rice cultivars revealed that the majority of local cultivars had late maturing, hence required prolonged and consistent water supply. The localities from where local cultivars were obtained are characterized by red, brown, and laterite soils of high rainfall areas (Dikshit 1986). The areas of scented rice cultivation are based on the foothills where temperatures are comparatively low, having typical soil conditions and moisture that could be taken as ideal conditions for production of high aroma. These regions naturally fulfill desirable conditions and hence are considered as niche areas for scented rice cultivation. Even though these conditions are common in Konkan region, we could collect only two cultivars from this region. Similarly, Bhandara and Gondia districts are known as rice bowls of Maharashtra (http://www.maharashtra.gov.in). In this region, landraces were not observed to be under cultivation. This could be attributed to genetic erosion by the high-yielding rice varieties provided by government agencies. Owing to higher economic returns, hybrid varieties have replaced vast number of scented landraces. As revealed by farmers, few decades ago several renowned scented rice landraces were cultivated in these regions, including Jirvel, Nagmoti, and Hiranakhi, which are nowhere to be found in cultivation.

Such genetic erosion has been reported by Patil and Patil (1994) in North Konkan region of Maharashtra where they reported that in 1982–1983 share of high-yielding varieties rapidly increased from 59.25 to 80.68 % in 1990–1991 and area under local varieties of rice during the corresponding period decreased from 40.75 to 19.32 %. Despite the encroachment of high-yielding varieties, landraces like Ambemohar and Ghansal survived the onslaught of high-yielding varieties owing to their characteristic aroma, taste, and stability of yield in their niche areas. They fetch 15–20 % high price than other varieties and are in demand in Pune and Mumbai markets. From Ambemohar, pure line selections Ambemohar-157, Ambemohar-159, and Ambemohar-102 were developed and released

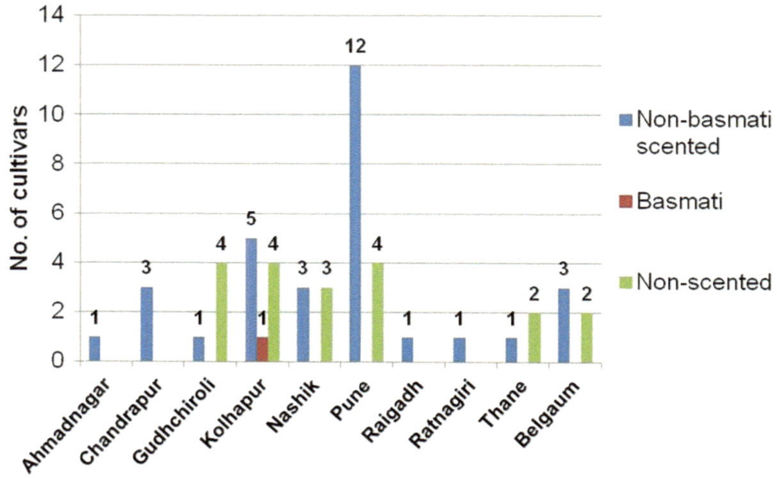

Fig. 2.2 District-wise details of field collection of rice (*O. sativa* L.) cultivars. The numerals over each bar indicate number of cultivars belonging to that category

for cultivation. However, at present only Ambemohar-157 is under cultivation. Similarly, Krishnasal from Ahmednagar district, Banaspatri, and Chinoor from Vidarbha region were released for cultivation; nevertheless, we came across only Chinoor types from this region. The reduction in cultivation area of landraces can also be attributed to water resource management like construction of dams. The niche area in villages like Talegaon near the Vaitarana Dam in Nashik district submerged in back water was once known for the cultivation of Kamod, scented landrace. As a result, Kolpi, a nonscented cultivar, is the sole cultivar grown in this village. During the study, we also came across various factors that are influencing the quality and quantity of aroma. As learned from farmers, the major affecting factor was the use of chemical fertilizers in efforts to increase production instead of the traditional farmyard manure. Similar observations were recorded by Suwanarit et al. (1996), where they found that aroma and other physicochemical characters in Khao Dawk Mali 105 are adversely affected by the applied dosages of nitrogen. Association of rhizosphere microbes have been reported with rice seedlings that can synthesize rice aroma when cultured in isolation on synthetic medium (Romanczyk et al. 1995, Rungsardthong and Noomhorm 1995, Nugsuk et al. 2003). Urea-based fertilizers are used by majority of farmers to improve yield. However, inefficiency of urea as fertilizers results in increase of pH leading to toxicity (Court et al. 1964). This might be disturbing the microbial flora and hence the aroma contents. Therefore, due to multiple attributes such as late maturity, low yield, mixing with other varieties, and reduction in aroma content, a large number of scented landraces have already been lost and many are at the verge of extinction.

2.3.1.2 Collection by Procurement from Research Stations and Institutes

Collection by procurement from research stations and institutes consisted of 62 nonbasmati scented, 7 basmati, and 2 nonscented rice cultivars. The details of cultivars are presented as Fig. 2.3 and Table 2.4.

ARRS, Radhanagari, and RRS, Shindevahi, possess impressive number of scented rice cultivars. Among these, Giriga sambha is native to Andhra Pradesh. Similarly, Bantaphool (Himachal Pradesh), Adamchini, Bantaphool, Kanakjeer, Kalanamak, dubraj (Uttar Pradesh), Shyamjeer, Badshahbhog, Kanakjeer, Kalanamak, Tulsimanjula, Gopalbhog (Bihar), Kalajeera (Assam), Badshahbhog (West Bengal), Lalu, Vishnubhog, dubraj, badshahbhog Tulsiamrit, adamchini, bantaphool (Madhya Pradesh), Kalajeera, Durgabhog, Pimpudibasa, Prabhatjeera Jhilipanjari Dubrajsena, Kalajeera,

Table 2.3 Details of field collection of rice (*O. sativa* L.) cultivars

Type	State	District	Category	Cultivar
Nonbasmati scented	Maharashtra	Ahmadnagar	Landrace	Kala bhat
		Chandrapur	Landrace	Kate chinoor, Kondhekar chinoor, Parabhani chinoor
		Gudhchiroli	Hybrid	Makarand
		Kolhapur	Landrace	Ambemohar Ajra, Champakali, Ghansal, Girga, Kothmirsal
		Nashik	Landrace	Chimansal, Kamod
			Hybrid	Indrayani
		Pune	Landrace	Ambemohar, Ambemohar Pandhara, Ambemohar-Tambda, Jiri, Kalsal, Kamavatya, Khadkya, Lal bhat, Lal dodki, Raibhog, Tamsal
			Selection	Ambemohar-157
		Raigadh	Landrace	Gham
		Ratnagiri	Landrace	Velchi
		Thane	Hybrid	Pusa sugandha
	Karnataka	Belgaum	Landrace	Basumati, Kali kumud, Kumud
Basmati	Maharashtra	Kolhapur	Hybrid	Basmati
Nonscented	Maharashtra	Gudhchiroli	Landrace	Luchai, Mothi luchai, Navedhan, Suvarna wild
		Kolhapur	Hybrid	Ratnagiri 24, Sahyadri, Sonam, Sonamasuri
		Nashik	Landrace	Dev bhat, Gari kolpi, Hali kolpi
		Pune	Landrace	Kolamb, Majeri, Pandhara dodki
		Pune	Hybrid	Mahsuri
		Thane	Landrace	Zeeni kolam
		Thane	Hybrid	Swarna
	Karnataka	Belgaum	Landrace	Chitak bhat, Manila

Kala Krishna, Acharmati, Gatia, Chinikamini (Orrisa) and Shrikamal, and Elaichi (Chattisgarh) are native to states other than Maharashtra and Karnataka (Singh et al. 2000a, Rohilla et al. 2000, Shobha Rani and Singh 2003, Das et al. 2003). Inclusion of these cultivars provides a wider scope for comparison among scented rice cultivars. Eight landraces (Ambemori, Gandhesale, Geerige sanna, Medhini sanna bhatta, Mysore mallige, Mysore sanna, Sanna bili bhatta, and Vasane sanna bhatta) procured from ARS, Sirsi, and two (Kagisali, Kaligajvili) from RRS, Mugad, are native to Karnataka and are under cultivation in various regions of Karnataka.

2.3.1.3 Procurement of Hybrid Basmati Cultivars

The seeds of hybrid basmati cultivar Pusa basmati-1 were procured from National Seed Corporation Ltd., New Delhi (NSC, New Delhi).

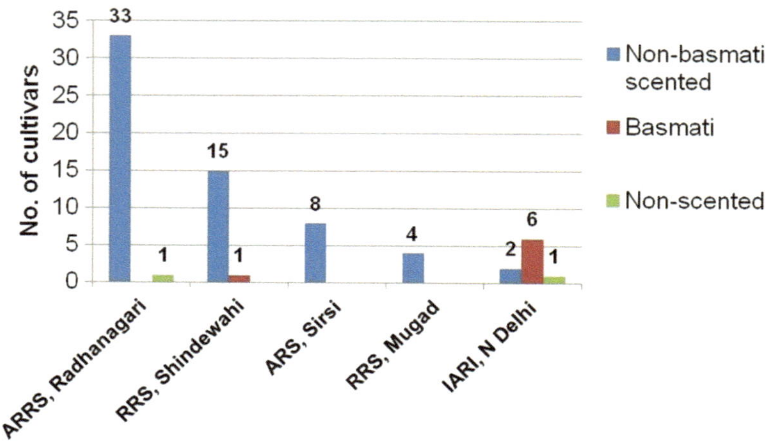

Fig. 2.3 Details of rice (*O. sativa* L.) cultivars procured from research stations. The numerals over each bar indicate the number of cultivars belonging to that category

The grains of selected rice cultivars from collection representing nonbasmati scented types, basmati types, and nonscented types are presented as Figs. 2.4, 2.5, and 2.6 respectively.

Characterization of Scented Rice Cultivars for Variability in Grain Morphology

Cultivars were characterized on the basis of awning (awn length and color), color and pubescence of lemma palea, sterile lemma color, and apiculus color (Table 2.4). As seen from Table 2.6, among 88 cultivars, awn was absent in 64 while other cultivars represent variants in awn size (short awn – 13 cultivars, long awn – 11 cultivars) and awn color (straw – 17 cultivars, red – 2 cultivars, and black – 5 cultivars). The color of lemma palea varied as straw (62), golden (5), red (6), brown (2), and black (13 cultivars). Cultivars also exhibited variation in the presence of furrows (gold, brown, and purple) with or without spots (brown and purple), and pubescence (53 cultivars) on lemma palea. The sterile lemma also recorded variation in color, viz., straw (55), gold (5), red (9), and purple (19 cultivars). Cultivars included variants for apiculus color, i.e., straw (44), red (3), purple (2), brown (12), and black (27). Variation in grain morphology of some cultivars is presented in Fig. 2.7.

Considerable variation in grain morphology was observed in cultivars procured from rice research stations, as well as in field collection. Cultivars represent variants in terms of absence or presence of varied colored and sized awn, lemma palea color with absence or presence of colored furrows and spots, along with pubescence and varied colored apiculus and sterile lemma. A study on scented rice cultivars from Asia indicated that morphological traits were useful for preliminary evaluation and could be used as general approach for assessing genetic diversity among morphologically distinguishable scented rice cultivars (Hien et al. 2007). Considering the extent of variation grain morphology, it could be used for the assessment of diversity as suggested by Hien et al. (2007). Patra and Dhua (2003) reported that among 120 cultivars of upland rice collected from Jaypore tract of Orrisa, very less variation in morphological characters was observed. In comparison to this, Maharashtra exhibited a wide range of variation in scented rice germplasm. Bansal et al. (2006) evaluated 666 lines of scented rice, including 520 of Indian origin. They recorded that in the germplasm, 51.1 % of accessions were awnless, 58.9 % were with straw color of lemma and palea, and apiculus color of straw and brown was predominant. As seen from Table 2.5, these characters were also found to be frequent among the cultivars assessed in the present study.

As seen from Table 2.6, majority of cultivars can be segregated on the basis of grain morphology. The variation recorded in the grain morphology of nonbasmati scented rice cultivars can be used as passport data for the identification of cultivars. Mathure et al. (2011) reported that

Table 2.4 Details of rice (*O. sativa* L.) cultivars procured from research stations

Type	State	Institute	Category	Cultivar
Nonbasmati scented	Maharashtra	ARRS, Radhanagari	Landrace	Adamchini-B, Amritbhog, Bansphool-A, Bantaphool-A, Barke bhat, Bela blue, Bishnubhog, Dhanprasad, Dubrajsena, Durgabhog, Gatia, Jeeraphool, Jhilipanjari, Kalakrishna, Kali kajari, Kanakjeer, Kothimbiri, Lalu, Pakhe bhat, Parbhatjira, Rati bhog, Shrabanmasi, Shyamjeer, Tulsi local, Tulsiganthi, Tulsikanthi, Velkat, Vishnubhog
			Selection	RDN local, RDN scented
			Hybrid	Bhogavati, Pawana, Phule radha
		RRS, Shindevahi	Landrace	Acharmati, Chinikamini, Dubraj, Dusara, Elaichi, Girija sambha, Gopalbhog, Jeera-sona, Kalajeera, Kalanamak, Pim-pudi-basa, Shrikamal, Shrikant, Tulshiamrit, Tulshimanjula
	Karnataka	ARS, Sirsi	Landrace	Ambemori, Gandhesale, Geerige sanna, Medhini sanna bhatta, Mysore mallige, Mysore sanna, Sanna bili bhatta, Vasane sanna bhatta
		RRS, Mugad	Landrace	Badshahbhog, Kagisali, Kaligajvili
			Hybrid	Mugad sugandha
	New Delhi	IARI, N Delhi	Selection	Kalanamak-3119, Kalanamak-3131
Basmati	Maharashtra	RRS, Shindevahi	Landrace	Pakistan basmati
	New Delhi	IARI, N Delhi	Landrace	Taraori basmati
			Selection	Basmati 370
			Hybrid	CSR-30, Pusa basmati, Pusa sugandha 5 (P-2511), Super basmati
Nonscented	Maharashtra	ARRS, Radhanagari	Hybrid	Sonsali
	New Delhi	IARI, N Delhi	Hybrid	Pusa 44

a

Fig. 2.4 (**a**) Selected nonbasmati scented type rice (*O. sativa* L.) cultivars from Maharashtra. (**b**) Selected nonbasmati scented type rice (*O. sativa* L.) cultivars from Karnataka. (**c**) Selected nonbasmati scented type rice (*O. sativa* L.) cultivars from other states of India

Fig. 2.4 (continued)

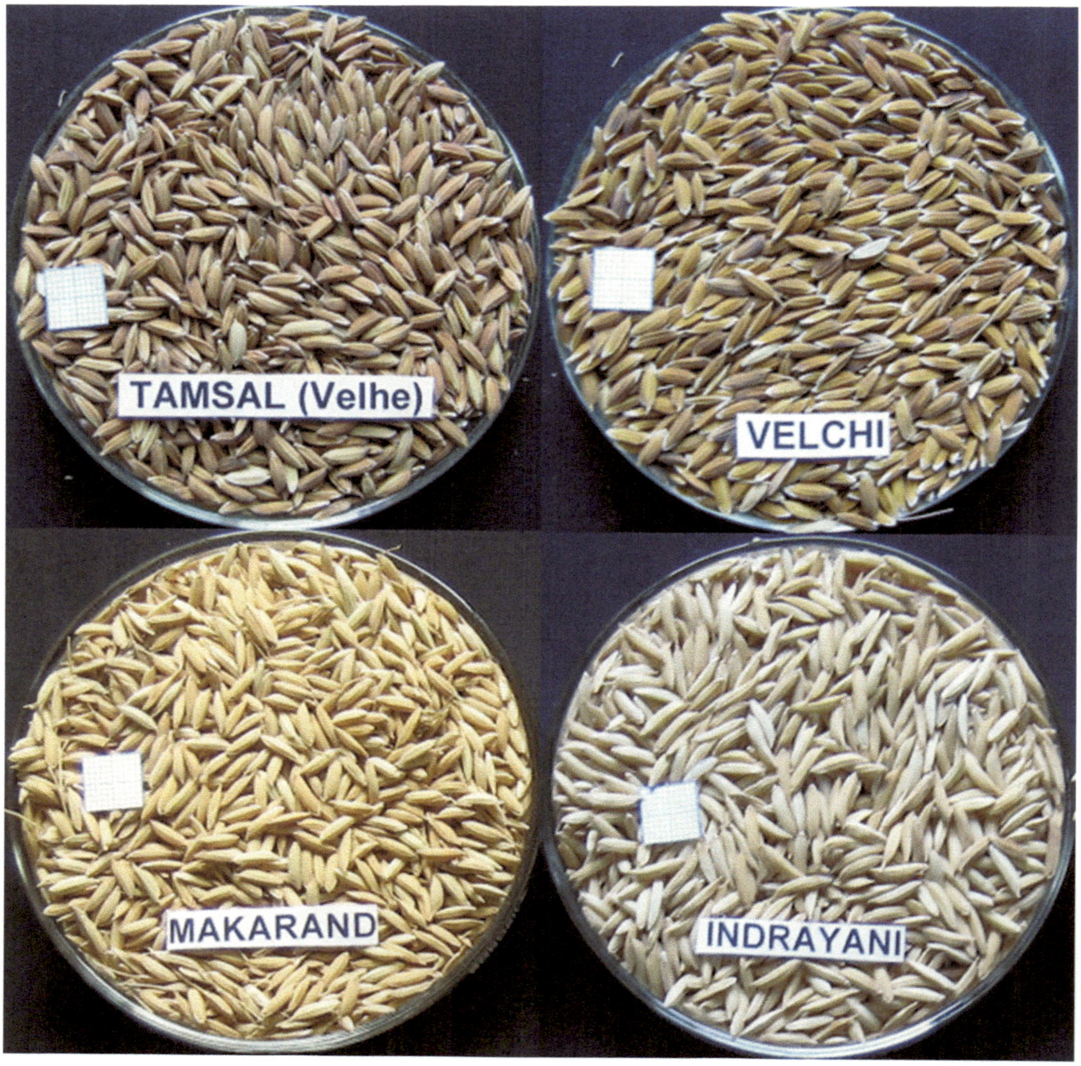

Fig. 2.4 (continued)

b

Fig. 2.4 (continued)

c

Fig. 2.4 (continued)

Fig. 2.4 (continued)

Fig. 2.4 (continued)

Fig. 2.4 (continued)

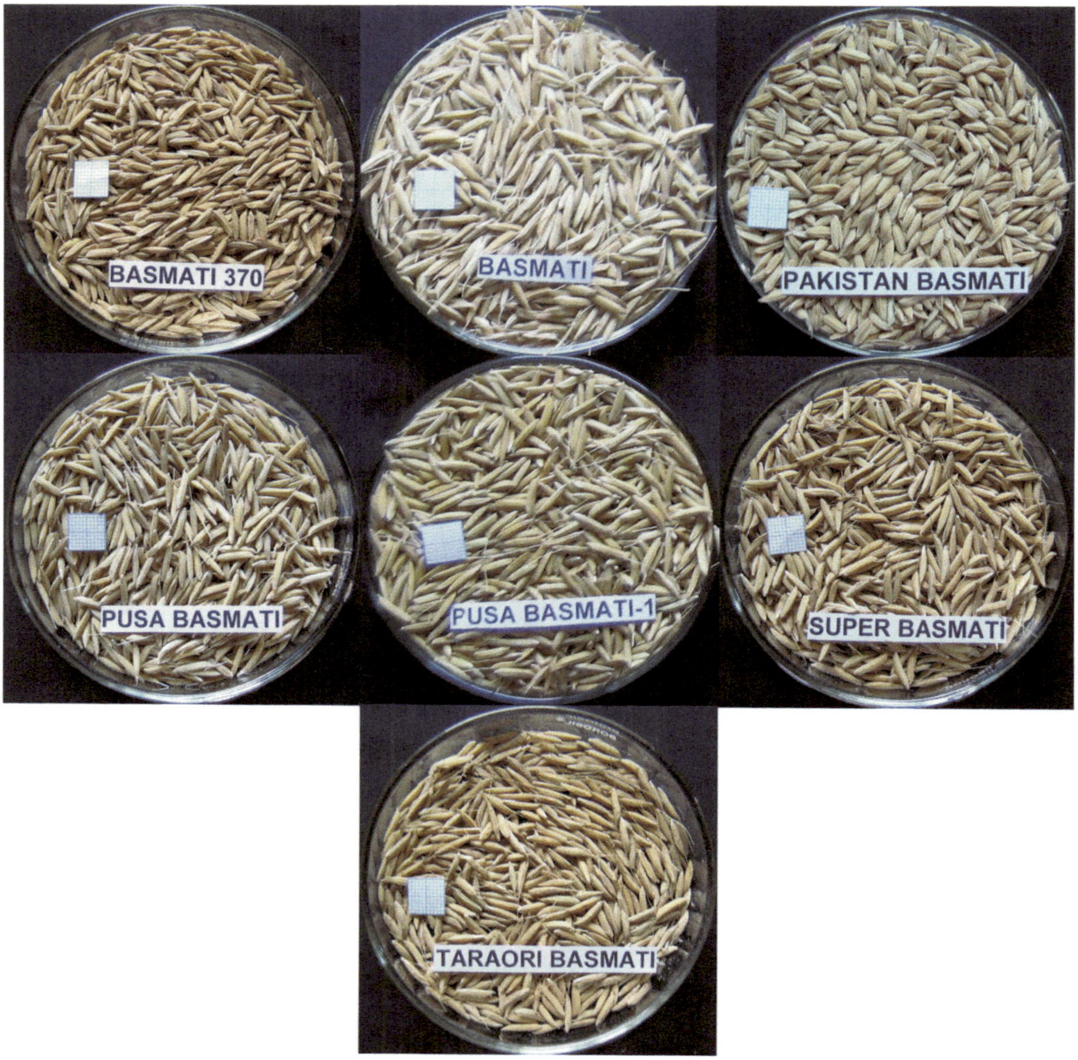

Fig. 2.5 Selected basmati type rice (*O. sativa* L.) cultivars

the lemma and palea morphology such as presence of awn and pubescence is associated with agromorphological traits.

They (Mathure et al. 2011) further indicated that presence of awn can be used as a selection parameter for slender fine-grained cultivars with higher number of productive tillers per plant and presence of pubescence on lemma palea can be utilized as a morphological marker owing to its positive association with kernel length, test weight, dwarf stature, and short panicle length.

Thus, characterization of cultivars with respect to grain morphology in the present study revealed encouraging potential of nonbasmati scented cultivars.

Fig. 2.6 Selected nonscented type rice (*O. sativa* L.) cultivars

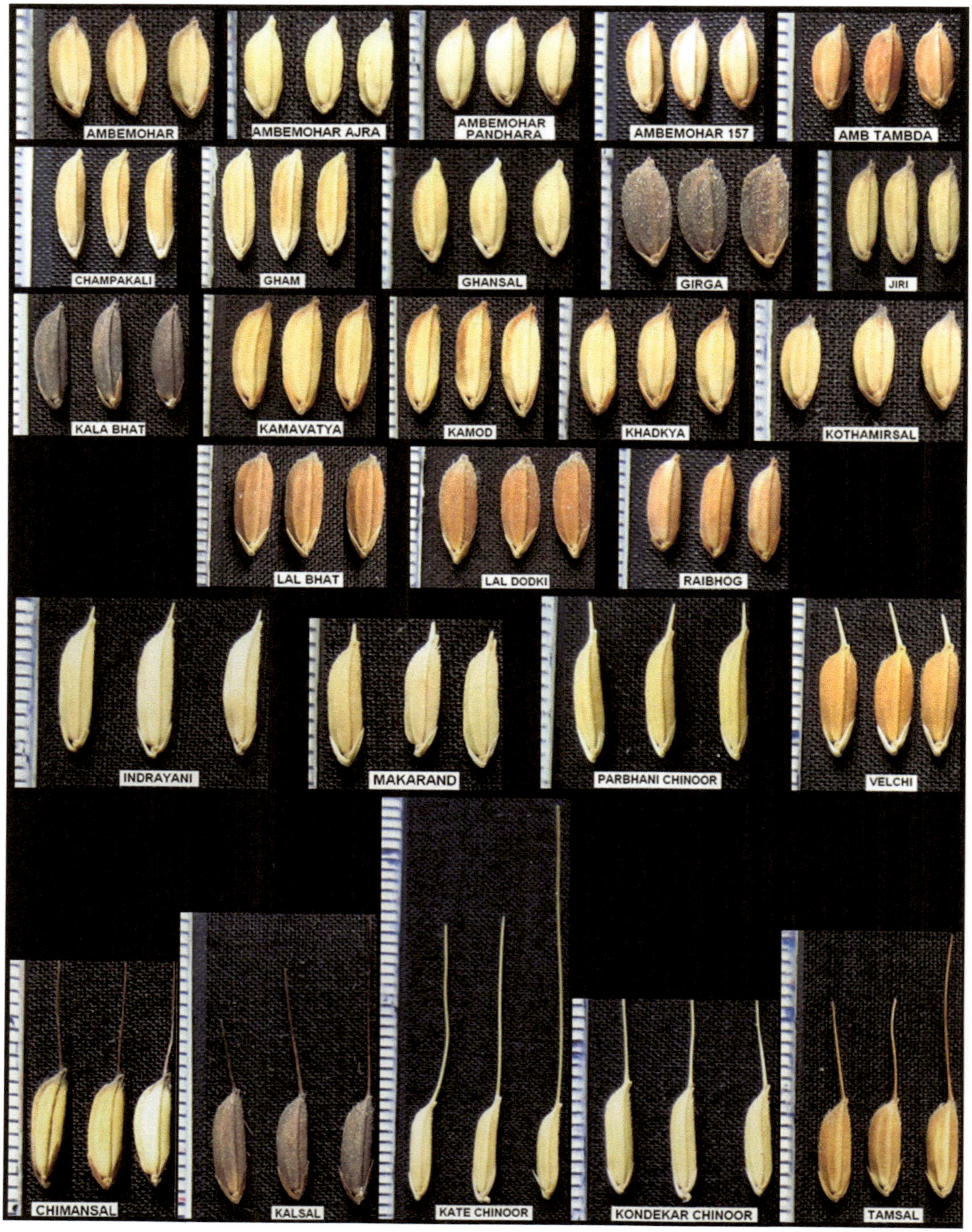

Fig. 2.7 Variation in grain morphology of selected scented rice (*O. sativa* L.) cultivars

Table 2.5 Characterization for grain morphology of 88 scented rice (*O. sativa* L.) cultivars

Cultivar	C1	C2	C3	C4	C5
Acharmati	0	6	1	1	1
Adamchini-B	0	1	1	4	7
Ambemohar	0	6	0	2	7
Ambemohar Ajra	0	1	0	1	1
Ambemohar Pandhara	0	4	0	2	6
Ambemohar-157	0	7	0	4	4
Ambemohar-Tambda	0	12	0	3	2
Amritbhog	0	5	0	1	1
Badshahbhog	0	4	0	2	6
Bansphool-A	0	1	1	1	1
Bantaphool-A	0	17	1	4	7
Barke bhat	0	15	1	1	1
Basmati	1	1	1	1	1
Basmati 370	4	1	1	1	1
Basumati	1	1	0	1	1
Bela blue	0	14	1	1	1
Bhogavati	0	1	1	1	1
Bishnubhog	0	1	1	1	1
Champakali	0	10	1	1	1
Chimansal	6	6	0	2	7
Chinikamini	0	1	1	1	1
CSR-30	4	1	0	1	1
Dhanprasad	0	1	0	1	7
Dubraj	4	5	1	1	1
Dubrajsena	0	1	0	1	1
Durgabhog	0	1	1	4	7
Dusara	0	1	1	1	1
Elaichi	0	1	0	1	1
Gatia	2	2	0	3	6
Gham	0	10	1	1	1
Ghansal	0	1	0	1	1
Girga	0	17	1	4	7
Girija sambha	0	6	0	4	7
Gopalbhog	0	5	0	1	1
Indrayani	1	1	1	1	1
Jeeraphool	0	4	0	1	1
Jeerasona	0	1	1	1	1
Jhilipanjari	0	6	0	4	3
Jiri	0	6	0	1	6
Kala bhat	0	16	1	4	7
Kalajeera	0	17	1	4	7
Kalakrishna	0	2	0	4	7
Kalanamak	0	16	1	2	7
Kalanamak-3119	0	16	1	4	7
Kalanamak-3131	0	16	1	4	7
Kali kumud	0	16	1	4	7
Kalsal	6	16	1	3	7
Kamavatya	0	6	0	3	6
Kamod	0	7	1	1	2
Kanakjeer	3	17	1	4	7
Kate chinoor	4	5	0	1	1
Khadkya	0	6	0	3	6
Kondhekar chinoor	4	4	1	1	1
Kothimbiri	0	5	1	1	5
Kothmirsal	0	6	1	1	7
Kumud	0	4	0	1	1
Lal bhat	0	13	1	3	7
Lal dodki	0	13	1	3	7
Lalu	0	13	1	1	2
Makarand	1	1	1	1	1
Pakhe bhat	3	2	1	4	7
Parabhani chinoor	1	1	0	1	1
Parbhatjira	0	8	1	1	1
Pawana	4	10	1	1	1
Pim-pudi-basa	0	4	0	1	1
Pusa basmati	1	1	1	1	1
Pusa basmati-1	1	1	1	1	1
Pusa sugandha	1	1	1	1	1
Pusa sugandha 5	0	1	1	1	1
Raibhog	0	12	0	3	6
Rati bhog	0	17	0	4	7
RDN local	0	1	1	1	1
RDN scented	0	11	1	1	6
Shrabanmasi	0	5	1	1	1
Shrikamal	0	6	0	4	7
Shrikant	0	3	1	4	7
Shyamjeer	3	17	1	1	7
Super basmati	4	1	0	1	1
Tamsal	5	13	1	3	6
Taraori basmati	4	1	1	1	1
Tulshiamrit	0	17	0	1	7
Tulshimanjula	0	8	0	1	6
Tulsiganthi	0	1	0	1	6
Tulsikanthi	0	1	1	4	7
Velchi	1	9	1	1	1
Velkat	0	1	1	1	1
Vishnubhog	0	1	0	1	1

C1: awing, C2: lemma and palea color, C3: pubescence on lemma and palea, C4: sterile lemma color, C5: apiculus color

Table 2.6 Frequency distribution of grain morphology among scented rice (*O. sativa* L.) cultivars

Character	Description	No of cultivars	Percent
Awning	Absent	64	72.7
	Short straw colored	9	10.2
	Short red colored	1	1.1
	Short black colored	3	3.4
	Long straw colored	8	9.1
	Long red colored	1	1.1
	Long black colored	2	2.3
Lemma and palea color	Straw	31	35.2
	Brown spots on straw	3	3.4
	Purple spots on straw	1	1.1
	Straw with furrows	7	8.0
	Straw with golden furrows	6	6.8
	Straw with brown furrows	10	11.4
	Straw with purple furrows	2	2.3
	Brown spots on straw with furrows	2	2.3
	Golden	1	1.1
	Golden with furrows	3	3.4
	Golden with brown furrows	1	1.1
	Red	2	2.3
	Red with furrows	4	4.5
	Brown	1	1.1
	Brown with furrows	1	1.1
	Black	6	6.8
	Black with furrows	7	8.0
Pubescence on lemma and palea	Absent	35	39.8
	Present	53	60.2
Sterile lemma color	Straw	55	62.5
	Gold	5	5.7
	Red	9	10.2
	Purple	19	21.6

(continued)

Table 2.6 (continued)

Character	Description	No of cultivars	Percent
Apiculus color	Straw	44	50.0
	Red	3	3.4
	Purple apex	1	1.1
	Purple	1	1.1
	Brown apex	1	1.1
	Brown	11	12.5
	Black	27	30.7

References

Bansal UK, Saini RG, Saxena S, Agarwal RC, Sharma AD (2006) Catalogue of aromatic rice (*Oryza sativa* L.) germplasm. Punjab Agricultural University, Ludhiana and National Bureau of Plant Genetic Resources, New Delhi

Bhatti MS (1998) Crop gene pools in the mountain areas of Pakistan and threats. In: Pratap T, Sthapit B (eds) Managing Agro biodiversity. ICIMOD, Kathmandu, pp 323–333

Chaudhary P, Gauchan D, Rana RB, Sthapit BR, Jarvis DI (2006) Potential loss of rice landraces from a Terai community in Nepal: a case study from Kachorwa, Bara. PGR Newsl 137:14–21

Court MN, Stephen RC, Waid JS (1964) Toxicity as a cause of the inefficiency of urea as a fertilizer. Eur J Soil Sci 15(1):42–48

Das SR, Roy JK, Kar M, Das S (2003) Aromatic rices of Orissa. In: Singh RK, Singh US (eds) A treatise on the scented rices of India. Kalyani Publishers, New Delhi, pp 355–375

Dikshit KR (1986) Maharashtra in maps. Maharashtra state board for literature and culture, Bombay, pp 31–42

Glaszmann JC (1987) Isozymes and classification of Asian rice varieties. Theor Appl Sci 74:21–30

Hanamaratti NG, Prashanthi SK, Salimath PM, Hanchinal RR, Mohankumar HD, Parameshwarappa KG, Raikar SD (2008) Traditional land races of rice in Karnataka reservoirs of valuable traits. Curr Sci 94(2):242–247

Hien NL, Sarhadi WA, Oikawa Y, Hirata Y (2007) Genetic diversity of morphological responses and the relationships among Asia aromatic rice (*Oryza sativa* L.) cultivars. Tropics 16(4):343–355

Hore DK (2005) Rice diversity collection, conservation and management in northeastern India. Genet Resour Crop Evol 52(8):1129–1140

IBPGR-IRRI Rice Advisory committee (1980) Descriptors for rice *Oryza sativa* L. International Rice Research Institute, Manila

Itani T (1993) History, cultivation and breeding of aromatic rice varieties. Bull Hiroshima Prefect Univ 5(1/2):267–281

Juma C (1989) The Gene Hunter Biotechnology and the scramble for seeds. Princeton University Press, Princeton, p 14

Khush GS (2000) Taxonomy and origin of rice. In: Singh RK, Singh US, Khush GS (eds) Aromatic rices. Oxford & IBH Pub. Co., New Delhi, pp 5–13

Mathure S, Shaikh A, Renuka N, Wakte K, Jawali N, Thengane R, Nadaf A (2011) Characterisation of aromatic rice (*Oryza sativa* L.) germplasm and correlation between their agronomic and quality traits. Euphytica 179(2):237–246

Nerkar YS, Patil VD, Misal MB, Katare NB (2003) Aromatic rices of Maharashtra. In: Singh RK, Singh US, Khush GS (eds) Aromatic rices. Oxford & IBH Pub. Co., New Delhi, pp 339–346

Nugsuk A, Winichphol N, Rugsardthong V (2003) Identification of 2-Acetyl-1-pyrroline, the principle aromatic rice flavour compound, in fungus cultures. In: The 2nd international conference on biodiversity and bioactive compound, 17–19 July 2003. Pattaya exhibition Center, Cholburi, Thailand

Patil HN, Patil VH (1994) An assessment of status of rice varieties in north konkan regions of Maharashtra. J Maharashtra Agric Univ 19(1):69–72

Patra BC, Dhua SR (2003) Agro-morphological diversity scenario in upland rice germplasm of Jeypore tract. Genet Resour Crop Evol 50(8):825–828

Rana JC, Negi KS, Wani SA, Saxena S, Pradheep K, Kak A. Pareek SK. Sofi PA (2009) Genetic resources of rice in the Western Himalayan region of India: current status. Genet Resour Crop Evol 56(7):963–973

Rohilla R, Singh VP, Singh US, Singh RK, Khush GS (2000) Crop husbandry and environmental factors affecting aroma and other quality traits. In: Singh RK, Singh US, Khush GS (eds) Aromatic rices. Oxford & IBH Pub. Co, New Delhi, pp 155–177

Romanczyk JLJ, McClelland CA, Post LS, Aitken WM (1995) Formation of 2-Acetyl-1-pyrroline by several Bacillus cereous strains isolated from cocoa fermentation boxes. J Agric Food Chem 43:469–475

Rungsardthong V, Noomhorm A (1995). Microbial Production of aromatic rice flavours, 2-Acetyl-1-pyrroline compounds. Thai Patent 13175

Shobharani N, Singh RK (2003) Efforts on aromatic rice improvement in India. In: Singh RK, Singh US (eds) A treatise on the scented rices of India. Kalyani Publishers, New Delhi, pp 23–72

Siddiq EA (1992) Rice in India – potential areas for research and development. Oryza 29:1–6

Singh RK, Singh US (2003) A treatise on the scented rices of India. Kalyani Publishers, New Delhi

Singh RK, Singh US, Khush GS, Rohilla R, Singh JP, Singh G, Shekhar KS (2000a) Small and medium grained scented rices of India. In: Singh RK, Singh US, Khush GS (eds) Aromatic rices. Oxford & IBH Pub. Co, New Delhi, pp 155–177

Singh RK, Gautam PL, Saxena S, Singh S (2000b) Scented rice germplasm: conservation, evaluation and utilization. In: Singh RK, Singh US, Khush GS (eds) Aromatic rices. Oxford & IBH Pub. Co, New Delhi, pp 107–133

Suwanarit A, Kreetapirom S, Buranakarn S, Varanyanond W, Tungtrakul P, Somboonpong S, Rattapat S, Ratanasupa S, Romyen P, Wattanapryapkul S, Naklang K, Rotjanakusal S, Pornurisnit P (1996) Effect of nitrogen fertilizer on grain qualities of Khao Dawk Mali-105. Kasetsart J (Nat Sci) 30:458–474

3 Quality Parameter Assessment in Scented Rice Cultivars

3.1 Introduction

Rice quality is a complex trait comprising many physicochemical characteristics. Consumers base their concept of quality on the grain's appearance, size and shape of the grain, behavior upon cooking, taste, tenderness, and flavor of cooked rice (Table 3.1). Appearance of milled rice is important to consumers. Thus, grain size and shape are the first criteria for rice quality that breeders consider in developing new varieties (Adair et al. 1966). Preferences for grain size and shape vary with consumers as some ethnic groups prefer short bold grains, while medium and long slender grains are preferred by others. The cooking and eating qualities of rice are valuable properties, especially in Asia, where it is the most important food. These qualities are largely determined by the properties of starch that makes up to 90 % of milled rice. Gelatinization temperature, amylose content, and gel consistency are the important starch properties, which influence cooking and eating characteristics. Gelatinization temperature determines cooking time of rice. Amylose content correlates negatively with taste panel scores for cohesiveness, tenderness, color, and gloss of rice and hence predicts the cooking and processing behavior of rice (Juliano 1979). The texture of rice varies on the basis of gel consistency in high amylose content rice. In general, rice with soft to medium gel consistency, intermediate amylose content, and intermediate gelatinization temperature is a preferred level for the consumers (Khush et al. 1979). However, the cooking quality preferences are known to vary within the country, within the ethnic groups, from one country to another, and within different geographical regions (Juliano et al. 1964). Extra elongation on cooking is a highly desirable trait in high-quality rice like basmati (Table 3.2, Fig. 3.1). Thus, these parameters play a crucial role in rice quality.

Almost all parts of India produce one or the other type of scented rice for human consumption. Scented rice has high premium value in national as well as international markets owing to unique aroma and quality. Scented rice having small to medium slender grain and aroma are locally preferred owing to excellent cooking and eating quality traits. In spite of being aware of the spectacular yields of modern varieties, some farmers still prefer to grow the low-yielding traditional varieties because of their good quality attributes. These local varieties are a very rich source of gene pool.

Sarma et al. (1990) studied the grain characteristics of 43 traditional scented rice varieties of Assam and reported wide variation in grain length (5.7–9.9 mm), breadth (1.8–3.0 mm), L/B ratio (2.44–4.33), and 1000 grain weight (8.44–25.48 g). Physicochemical properties of seven scented cultivars, including basmati and 13 nonscented cultivars, were analyzed by Dutta et al. (1998). Four basmati type varieties of Punjab were assessed for superior quality characteristics by Sharma (2002).

A. Nadaf et al., *Scented rice (Oryza sativa L.) Cultivars of India: A Perspective on Quality and Diversity*, DOI 10.1007/978-81-322-2665-9_3

Table 3.1 Grain quality parameters in rice

Quality parameter	Character	Character state
[a]Physical characteristics	Grain size	Extra long (>7.5 mm),
		Long (6.61–7.5 mm),
		Medium (5.51–6.60 mm)
		Short (<5.5 mm)
	Grain shape	Slender (>3.0),
		Medium (2.1–3.0),
		Bold (1.1–2.0)
		Round (<1.0)
	Grain quality (based on test weight of 1000 grains)	Coarse (>25 g.),
		Medium (20–25 g),
		Fine (15–20 g)
		Superfine (<15 g)
[b]Chemical characteristics	Amylose content	Waxy (0–2 %),
		Very low (3–9 %),
		Intermediate (20–25 %)
		High (>25 %)
	Gel consistency	Hard gel consistency
		(gel length 40 or <40 mm),
		Medium gel consistency
		(41–60-mm gel length),
		Soft gel consistency
		(gel length >60 mm)
	Alkali spreading value	High (>74 °C)
		Intermediate (70–74 °C)
		Low (55–69 °C)
Cooking quality	Cooking coefficient	–
	Coefficient of elongation	
	Coefficient of latitudinal expansion	

[a]Cruz and Khush (2004), Shobharani et al. (2004)
[b]Juliano (1971), Cagampang et al. (1973), Little et al. (1958)

Table 3.2 Grain quality characteristics of basmati rice

Origin	Northern India at the foothills of the Himalayas
Color	Translucent, creamy white
Grain	Long grain (6.61–7.5 mm) or very long (more than 7.50-mm and 2-mm breadth)
Shape	Length-to-width ratio over 3.0
Texture	Dry, firm, separate grains; upon cooking, the texture is firm and tender without splitting, and it is nonsticky
Elongation	The rice elongates almost twice upon cooking but does not fatten much. When cooked, the grains elongate (70–120 % over the precooked grain) more than other varieties.
Flavor	Distinctive fragrance from a cocktail of 100 compounds – hydrocarbons, alcohols, aldehydes, and esters. A particular molecule of note is 2-acetyl-1-pyrroline having popcorn-like odor.

Dipti et al. (2002) studied the physicochemical and cooking properties of six fine rice varieties and recorded L/B ratio and amylose content in the range of 1.3–3.6 and 18.6–28 % respectively. Ten Beruin rice varieties studied by Dipti et al. (2003) were of medium-bold size and shape and of low amylose content. In 2003, Khatun et al. analyzed 16 scented rice varieties for 12 physicochemical properties and performed correlation studies. Kumar et al. (2003) recorded significant interaction between germplasm and environment with respect to the physical characteristics of seven scented rice varieties. A study to evaluate the performance of 12 promising rice varieties was undertaken in Kerala, India, by George et al. (2005). Physicochemical, cooking, and textural properties of 23 Indian rice cultivars and correlation between them were analyzed by Singh et al. (2005). In 2007, Sarawgi and Bisne conducted a divergence study on scented rice for agromorphological and quality traits to assist the selection of parent in hybridization. Popped rice is a popular snack in India; hence, Bhatupadya et al. (2008) assessed physicochemical characterization with respect to popping. A detailed study on Badshahbhog by Bisne and Sarawgi (2008) yielded eight genotypes for good grain quality with higher yield. Considering the splendid diversity of scented

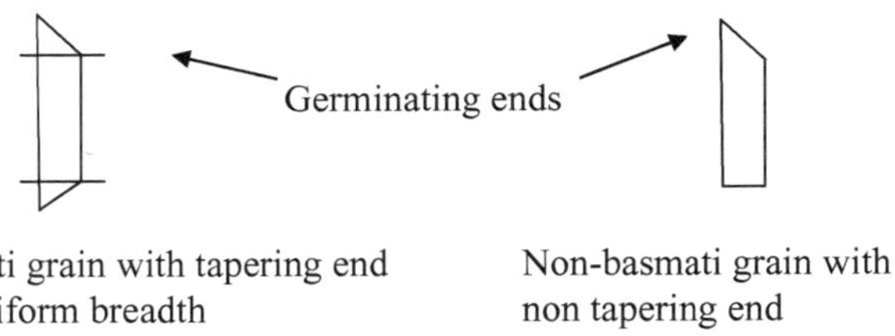

Fig. 3.1 Characteristics of basmati rice kernel (Adopted from Mahindru 1995)

rice in India, only few studies focused on scented landraces are available (Sarma et al. 1990; Sharma 2002; Kumar et al. 2003; George et al. 2005; Singh et al. 2005, Sarawgi and Bisne 2007, Bhatupadya et al. 2008).

As the information on the physicochemical and cooking characteristics of traditional scented varieties of Maharashtra and Karnataka is scanty, the present study was undertaken to characterize the scented rice cultivars with respect to their physicochemical properties and cooking characteristics.

3.2 Methodology and Approaches Followed for Grain Quality Determination

3.2.1 Rice Samples

Sixty-four cultivars containing 57 nonbasmati scented, 3 basmati, and 4 nonscented cultivars were assessed for physicochemical and cooking characteristics. The details of cultivars are given in Table 3.3.

3.2.2 Determination of Physical Characteristics

The appearance of rice is important to consumers. Grain appearance depends on size and shape of the kernel.

3.2.2.1 Determination of Grain Size and Shape

Twenty grains of each cultivar were dehusked manually and measured for length and breadth of the kernel. On the basis of kernel length, cultivars were classified into size categories following the classification of Cruz and Khush (2004). They are as follows: extra long (>7.5 mm), long (6.61–7.5 mm), medium (5.51–6.60 mm), and short (<5.5 mm). The grain shape was determined by taking L/B ratio. On the basis of shape, cultivars were classified as slender (>3.0), medium (2.1–3.0), bold (1.1–2.0), and round (<1.0) grain type (Cruz and Khush 2004).

3.2.2.2 Determination of Grain Quality

Twenty-five grains of each cultivar were weighed in triplicate, and 1000 grain weight (test weight) was calculated. On the basis of test weight, cultivars were classified as of coarse (TW >25 g.), medium (20–25 g), fine (15–20 g), and superfine (TW <15 g) grain quality (Shobharani et al. 2004).

3.2.3 Determination of Chemical Properties

3.2.3.1 Determination of Apparent Amylose Content

The apparent amylose content was determined following (Juliano 1971) with some modifications. About 30–35 grains were polished manually and ground to fine powder using mortar and pestle. The powder was passed through sieve of 50 mesh size to maintain smaller particle size. Rice powder measuring 100 mg was weighed on digital weighing balance and placed in 100-ml volumetric flask. A total of 1 ml of 95 % ethanol and 9 ml of 1 N NaOH were added to it. The content was heated in boiling water bath for 10 min to gelatinize the starch and was allowed to cool for 1 h at room temperature. After cooling, volume was made up to 100 ml with distilled water (DW) and contents are mixed well. In a 5-ml starch solution, one drop of 0.1 % phenolphthalein indicator was added, which turns pink. Using 1 N acetic acid, pH was adjusted to colorless. An amount of 2 ml of Iodine solution (0.2 g I_2+2 g KI in 100-ml aq. solution) was added, and vol-

Table 3.3 Details of rice (*O. sativa* L.) cultivars selected for assessment of physicochemical and cooking characteristics

Type	State	Cultivar
Nonbasmati scented	Maharashtra	Ambemohar, Ambemohar Ajra, Ambemohar Pandhara, Ambemohar-Tambda, Champakali, Chimansal, Gham, Ghansal, Girga, Jiri, Kala bhat, Kalsal, Kamavatya, Kamod, Kate chinoor, Khadkya, Kondhekar chinoor, Kothmirsal, Lal dodki, Lal bhat, Parabhani chinoor, Raibhog, Tamsal, Velchi, Ambemohar-157, Indrayani
	Karnataka	Ambemori, Basumati, Gandhesale, Geerige sanna, Kagisali, Kaligajvili, Kali kumud, Kumud, Medhini sanna bhatta, Mysore mallige, Mysore sanna, Sanna bili bhatta, Vasane sanna bhatta
	Other states	Acharmati, Badshahbhog, Barke bhat, Bela blue, Dubraj, Dusara, Elaichi, Jeera-sona, Kalanamak, Kali kajari, Kothimbiri, Pakhe bhat, Ratibhog, Shrikamal, Shrikant, Tulshimanjula, Tulsi local, Velkat
Basmati		Pakistan basmati, Basmati 370, Pusa basmati-1
Nonscented		Kolamb, Chitak bhat, Manila, Sahyadri

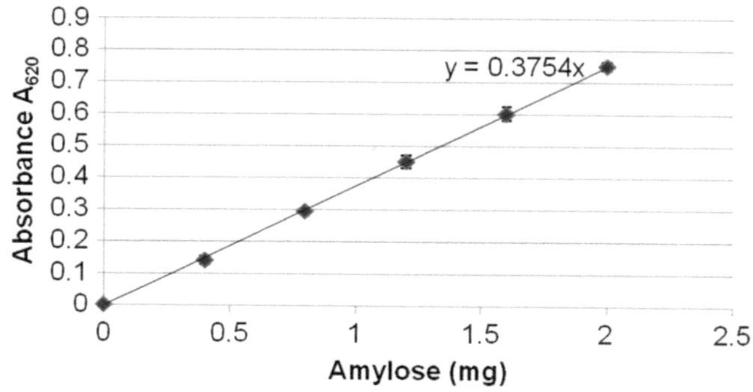

Fig. 3.2 Standard graph developed using standard amylose

ume was made up to 100 ml with DW. The content was allowed to stand for 20 min. Absorbance of the solution was measured at 620 nm.

Development of Standard Graph for Amylose Content

For the standard curve, 40-mg amylose (Sigma-Aldrich Inc.) was wetted with 1 ml of 95 % ethanol and 9 ml of 1 N NaOH and heated for 10 min in a boiling water bath, cooled and made up the volume to 100 ml. Amounts of 1, 2, 3, 4, and 5 ml of this solution (resulting amylose concentrations 0, 0.4, 0.8, 1.2, 1.6, and 2 mg) were placed in 100-ml volumetric flasks, and one drop of 0.1 % phenolphthalein indicator was added. The content was acidified with acetic acid and treated as mentioned earlier. Absorbance of the solutions was measured at 620 nm using a spectrophotometer. A standard graph was developed by plotting absorbance values against concentration of amylose (mg). Trend line with linear correlation coefficient approaching 1 was drawn to determine the conversion factor (Fig. 3.2). The dilution factor of 20 for the samples was included in the conversion factor. Apparent amylose content of each sample was determined by the following formula:

$$\text{Apparent amylose content (AC)}\% = (A_{620} / 0.3754) \times 20$$

where

A_{620} = absorbance at 620 nm,
0.3754 = conversion factor based on standard graph,
20 = dilution factor.

The results were expressed as percent apparent amylose content in milled rice weight. Rice varieties were grouped on the basis of their amylose content into waxy (0–2 %), very low (3–9 %), intermediate (20–25 %), and high (>25 %) (Kumar and Khush 1986).

3.2.3.2 Determination of Gel Consistency

The gel consistency was determined following the method of Cagampang et al. (1973). Rice grains were polished manually and ground to fine powder. It was sieved through 100-mesh size sieve. An amount of 100-mg rice powder was weighed in duplicates into 13-×100-mm culture tubes and wetted with 0.2-ml 95 % ethanol with 0.03 % thymol blue. Exactly 2-ml 0.2 N KOH was added to it, and content was mixed well for 2–3 s. Tubes were covered with glass marbles and heated in boiling water bath for 8 min. It was ensured that the tube contents reach two thirds the height of the tube. The tubes were allowed to stand for 5 min at RT and cooled in ice water bath for 20 min and laid horizontally on millimeter graph paper for 1 h. Length of blue colored gel was measured in the bottom of the tube to the gel front. The varieties under study were classified based on gel length as follows: very flaky rice with hard gel consistency (length of gel 40 or <40 mm), flaky rice with medium gel consistency (41–60-mm gel length), soft rice with soft gel consistency (length of gel >60 mm).

3.2.3.3 Determination of Gelatinization Temperature (by Alkali Spreading Value)

The time required for cooking is determined by gelatinization temperature of starch. Gelatinization temperature is the range of temperature wherein at least 90 % of the starch granules swell irreversibly in hot water. The gelatinization temperature was determined following alkali digestibility test (Little et al. 1958). Six whole grains without cracks were placed in Petri plates containing 20-ml 1.7 % KOH. The Petri plates were covered and incubated at 30 °C for 23 h, and starchy endosperm was rated visually based on a seven-point numerical spreading scale (Table 3.4). As control, six whole grains without cracks were placed in DW and assessed as test samples.

3.2.4 Determination of Cooking Characteristics

Bran layer of 40 grains was removed manually. Grains were cooked in excess of water in a beaker on heating mantle. To ensure complete cooking, grains were pressed in two slides periodically till there is no white core of endosperm. When three consecutive grains assessed had no white core, cooking was stopped. The cooked rice was transferred to blotting paper. Ten cooked whole grains were selected and measured for length and breadth using vernier caliper.

On the basis of grain length and breadth of cooked and uncooked grain, the coefficient of cooking, elongation, and latitudinal expansions were determined using the following formulae:

Cooking coefficient = (Lc – Bc)/(Lu – Bu)
Coefficient of elongation = Lc/Lu
Coefficient of latitudinal expansion = Bc/Bu
L/B ratio of cooked kernel = Lc/Bc

where

Lu = length of uncooked grain
Bu = breadth of uncooked grain
Lc = length of cooked grain
Bc = breadth of cooked grain

Table 3.4 A seven-point numerical spreading scale used to determine alkali spreading value

Score	Spreading pattern	Alkali digestion	Gelatinization temperature
1	Kernel not affected	Low	High (>74 °C)
2	Kernel swollen	Low	High (>74 °C)
3	Kernel swollen, collar complete or narrow	Low or intermediate	High (>74 °C) or intermediate (70–74 °C)
4	Kernel swollen, collar complete and wide	Intermediate	Intermediate (70–74 °C)
5	Kernel split or segregated, collar complete and wide	Intermediate	Intermediate (70–74 °C)
6	Kernel dispersed, merging with collar	High	Low (55–69 °C)
7	Kernel completely dispersed and intermingled	High	Low (55–69 °C)

3.3 Grain Quality Determination

3.3.1 Determination of Physical Characteristics

The determination of physical characteristics with respect to grain size, shape, and quality for the varieties under study is given bellow

3.3.1.1 Determination of Grain Size and Shape

Average grain length varied from 3.89 (Kothimbiri) to 7.17 mm (Sanna bili bhatta) in nonbasmati scented rice cultivars, 6.07 (Pakistan basmati) to 8.17 mm (Pusa basmati-1) in basmati, and 6.15 (Kolamb) to 7.4 mm (Sahyadri) in nonscented rice cultivars (Table 3.5). Among nonbasmati scented type, grain length varied from 4.09 (Ambemohar Ajra) to 7.03 mm (Gham) in cultivars from Maharashtra, 4.13 (Vasane sanna bhatta) to 7.17 mm (Sanna bili bhatta) in cultivars from Karnataka, and 3.89 (Kothimbiri) to 6.84 mm (Kali kajari) in cultivars from other Indian states. As seen from Fig. 3.3, majority of nonbasmati scented rice cultivars had short to medium grain size. Basmati cultivars were classified as of very long, long, and medium grain size. Among non-scented type, three cultivars were of long grain size and one cultivar of medium grain size.

Average grain breadth among cultivars varied from 1.59 mm (Parabhani chinoor) to 2.94 mm (Bela blue) in nonbasmati scented rice cultivars, 1.90 mm (Basmati 370) to 2.23 mm (Pakistan basmati) in basmati, and 2.29 mm (Kolamb) to 2.83 mm (Chitak bhat) in nonscented rice cultivars (Table 3.5). Among nonbasmati scented type, grain breadth varied from 1.59 (Parabhani chinoor) to 2.77 mm (Lal bhat), 2.07 (Geerige sanna) to 2.76 mm (Mysore sanna), and 1.82 (Kalanamak) to 2.94 mm (Bela blue) in cultivars from Maharashtra, Karnataka, and other Indian states respectively. Cultivars exhibited higher percent coefficient of variation (CV %) for grain length (18.64) than grain breadth (13.11).

L/B ratio of grain among rice cultivars varied from 1.4 (Kothimbiri) to 4.03 (Kate chinoor) in nonbasmati scented type, 2.73 (Pakistan basmati) to 4.28 (Pusa basmati-1) in basmati type, and 2.41 (Chitak bhat) to 3.19 (Sahyadri) in nonscented type (Table 3.3). Among nonbasmati scented type, state-wise, L/B ratio varied from 1.86 (Ambemohar-Tambda) to 4.03 (Kate chinoor) in cultivars from Maharashtra, 1.63 (Vasane sanna bhatta) to 3.24 (Basumati) in cultivars from Karnataka, and 1.4 (Kothimbiri) to 3.36 (Kali kajari) in cultivars from other Indian states. As seen from Fig. 3.4, medium grain shape was predominant (37 cultivars) over bold (11 cultivars) and slender (9 cultivars) shape among nonbasmati scented rice cultivars.

Basmati and nonscented cultivars were classified as of medium and slender grain shape. Among nonbasmati scented rice cultivars, Champakali, Gham, Kate chinoor, Kondhekar chinoor, Parabhani chinoor, and Indrayani from Maharashtra; Basumati and Kaligajvili from Karnataka; and Kali kajari from other states of India recorded slender grain shape.

There are few reports available on the traditional scented rice cultivars of Maharashtra. Ramarathanam and Kulkarni (1988) have reported L/B ratio of 2 in Ambemohar. Detailed study by Bisne and Sarawgi (2008) has recorded large variation in length (3.7–6.8 mm), breadth (1.2–2.3 mm), and L/B ratio (1.82–3.25) of Badshahbhog grains.

Grain dimensions observed for Gandhasale, Kumud, Kaligajvili, and Kagisali are comparable with the values reported by Mahadevappa and Mohankumar (2003). Earlier reports on grain dimensions of Basmati 370 vary from 6.34 to 6.97 mm for grain length, 1.84 to 1.92 mm for breadth, and 3.3 to 4.12 for L/B ratio (Juliano and Villareal 1993; Sharma 2002; Singh et al. 2005; George et al. 2005; Shobharani et al. 2006). Values observed for Basmati 370 during the present study are comparable with earlier reports.

Among the nonbasmati scented cultivars assessed, Basumati, Gham, Kaligajvili, Kali kajari, Champakali, Kate chinoor, Kondhekar chinoor, and Parabhani chinoor were classified under superior grain size and shape ranging from long slender to medium slender.

3.3.1.2 Determination of Grain Quality

Among rice cultivars, test weight varied from 7.33 (Jeera-sona) to 26.13 g (Lal bhat) in nonbasmati

Table 3.5 Determination of size, shape, and grain quality in rice (*O. sativa* L.) cultivars under study

Cultivar	Av. length (mm)	Av. breadth (mm)	L/B	Av. test weight (g)
Ambemohar	4.73±0.13	2.37±0.05	2.00	13.47±0.23
Ambemohar Ajra	4.09±0.17	1.78±0.06	2.30	8.00±0.40
Ambemohar Pandhara	4.79±0.22	2.30±0.06	2.08	13.47±0.46
Ambemohar-Tambda	4.98±0.11	2.67±0.09	1.86	13.07±0.23
Champakali	6.05±0.22	2.00±0.00	3.03	13.47±0.46
Chimansal	5.96±0.18	2.08±0.13	2.87	18.13±0.23
Gham	7.03±0.16	2.18±0.14	3.22	18.93±0.23
Ghansal	4.62±0.11	2.11±0.15	2.19	10.53±0.23
Girga	4.66±0.14	2.31±0.08	2.02	13.60±0.00
Jiri	4.90±0.31	2.07±0.09	2.37	11.47±0.23
Kala bhat	5.95±0.16	2.02±0.07	2.95	16.53±0.23
Kalsal	6.19±0.17	2.32±0.07	2.67	17.07±0.23
Kamavatya	5.60±0.17	2.03±0.05	2.76	15.07±0.23
Kamod	6.02±0.12	2.35±0.07	2.56	17.20±0.40
Kate chinoor	6.45±0.13	1.60±0.03	4.03	12.13±0.23
Khadkya	5.88±0.16	2.13±0.15	2.76	15.20±0.40
Kondhekar chinoor	6.31±0.19	1.68±0.04	3.75	11.07±0.46
Kothmirsal	4.31±0.16	2.06±0.12	2.09	14.00±0.40
Lal dodki	6.24±0.25	2.66±0.10	2.35	25.33±0.23
Lal bhat	6.50±0.16	2.77±0.12	2.35	26.13±0.23
Parabhani chinoor	6.07±0.17	1.59±0.06	3.81	11.47±0.23
Raibhog	4.85±0.18	2.27±0.05	2.13	15.73±0.46
Tamsal	6.07±0.15	2.25±0.16	2.70	18.13±0.23
Velchi	5.63±0.12	2.22±0.09	2.54	14.93±0.23
Ambemohar-157	4.93±0.16	2.47±0.07	2.00	13.47±0.23
Indrayani	6.90±0.16	2.00±0.00	3.45	17.73±0.23
Ambemori	4.39±0.07	2.39±0.04	1.84	11.87±0.23
Basumati	6.71±0.08	2.07±0.10	3.24	16.40±0.69
Gandhesale	4.67±0.07	2.16±0.05	2.16	10.80±0.00
Geerige sanna	4.40±0.05	2.07±0.05	2.13	9.07±0.23
Kagisali	6.13±0.07	2.54±0.05	2.41	18.93±0.61
Kaligajvili	6.87±0.28	2.24±0.06	3.07	16.93±0.23
Kali kumud	4.56±0.05	2.08±0.04	2.19	9.73±0.23
Kumud	4.68±0.10	2.21±0.06	2.12	9.60±0.40
Medhini sanna bhatta	4.80±0.19	2.41±0.08	1.99	10.13±0.23
Mysore mallige	6.14±0.17	2.35±0.05	2.62	17.87±0.23
Mysore sanna	4.58±0.07	2.76±0.06	1.66	16.00±0.40
Sanna bili bhatta	7.17±0.23	2.66±0.07	2.70	20.67±0.61
Vasane sanna bhatta	4.13±0.05	2.54±0.06	1.63	10.00±0.00
Acharmati	4.08±0.05	2.46±0.05	1.66	13.20±0.00
Badshahbhog	4.29±0.07	2.13±0.06	2.01	9.33±0.23
Barke bhat	6.05±0.11	2.58±0.04	2.34	22.13±0.23
Bela blue	5.84±0.22	2.94±0.14	1.99	25.47±0.46
Dubraj	5.87±0.15	1.99±0.10	2.95	13.20±0.40
Dusara	4.50±0.07	1.86±0.09	2.43	7.87±0.23
Elaichi	4.23±0.11	2.15±0.07	1.97	9.07±0.23

(continued)

Table 3.5 (continued)

Cultivar	Av. length (mm)	Av. breadth (mm)	L/B	Av. test weight (g)
Jeera-sona	4.43±0.13	1.93±0.11	2.30	7.33±0.23
Kalanamak	5.42±0.10	1.82±0.05	2.98	11.33±0.23
Kali kajari	6.84±0.17	2.02±0.04	3.38	17.87±0.23
Kothimbiri	3.89±0.09	2.78±0.04	1.40	16.13±0.23
Pakhe bhat	6.54±0.13	2.31±0.02	2.84	19.47±0.61
Ratibhog	4.85±0.18	2.27±0.05	2.13	12.27±0.23
Shrikamal	4.26±0.15	2.21±0.10	1.93	9.60±0.00
Shrikant	4.18±0.08	2.08±0.06	2.01	9.60±0.00
Tulshimanjula	4.42±0.09	2.04±0.11	2.17	8.40±0.00
Tulsi local	5.50±0.11	2.33±0.1	2.36	14.80±0.40
Velkat	5.70±0.13	2.36±0.1	2.42	13.73±0.23
Kolamb	6.15±0.23	2.29±0.1	2.68	17.73±0.61
Chitak bhat	6.81±0.21	2.82±0.1	2.41	24.27±0.46
Manila	6.72±0.22	2.45±0.1	2.75	19.47±0.23
Sahyadri	7.40±0.10	2.32±0	3.19	22.60±0.12
Pakistan basmati	6.07±0.17	2.23±0.1	2.73	18.00±0.40
Basmati 370	7.03±0.17	1.90±0.1	3.70	14.53±0.23
Pusa basmati-1	8.17±0.41	1.91±0.1	4.28	17.60±0.00
Average	5.53	2.23	2.53	14.82
CV %	18.64	13.11	24.15	31.04

±SD, Av.: Average

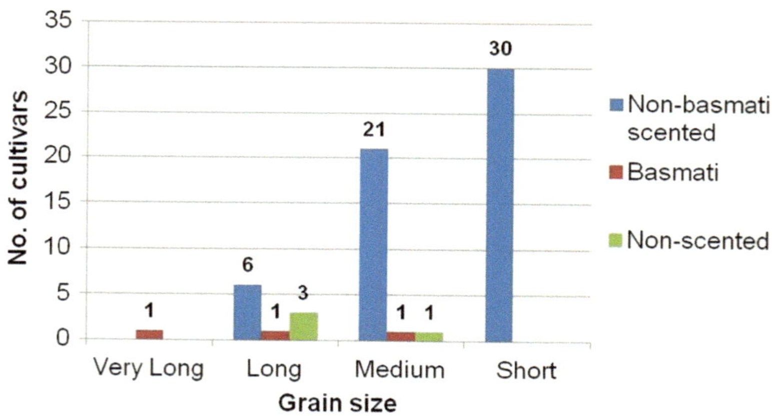

Fig. 3.3 *Classification of rice varieties under study on the basis of grain size.* The numerals over each bar indicate the number of cultivars belonging to that category

scented type, 14.53 (Basmati 370) to 18 g (Pakistan basmati) in basmati type, and 17.73 (Kolamb) to 24.27 g (Chitak bhat) in nonscented type rice cultivars (Table 3.3). Within nonbasmati scented type, test weight varied from 8 (Ambemohar Ajra) to 26.13 g (Lal bhat) in cultivars from Maharashtra, 9.07 (Geerige sanna) to 20.67 g (Sanna bili bhatta) in cultivars from Karnataka, and 7.33 (Jeera-sona) to 25.47 g (Bela blue) in cultivars from other Indian states. As seen from Fig. 3.5, majority of nonbasmati scented rice cultivars were of superfine (34 cultivars) and fine (18 cultivars) grain quality. Basmati cultivars were of superfine and fine quality, while nonscented cultivars were of fine and medium grain quality.

Test weight of Badshahbhog is reported as 9.2 g (Dutta et al. 1998) and 10.4 g (Shobharani and Singh 2003). Earlier reports on test weight of

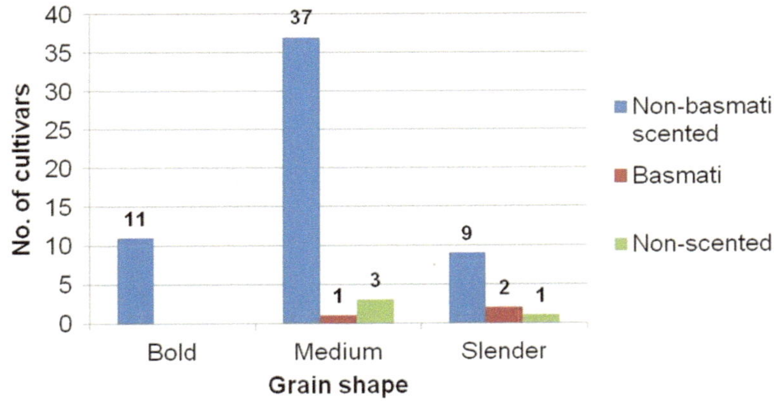

Fig. 3.4 *Classification of rice varieties under study on the basis of grain shape*. The numerals over each bar indicate the number of cultivars belonging to that category

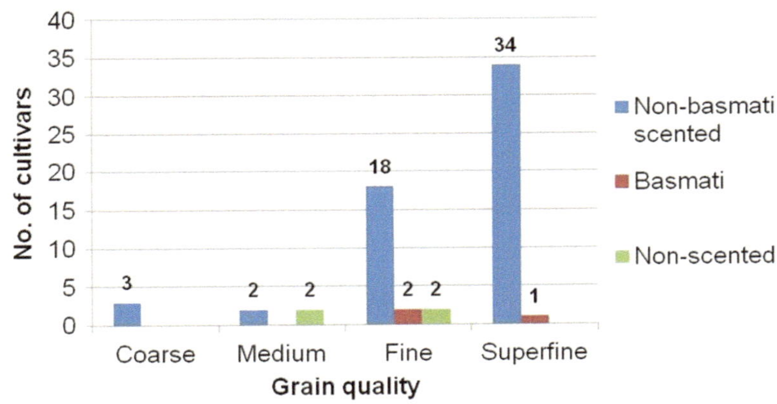

Fig. 3.5 *Classification of rice varieties under study on the basis of grain quality*. The numerals over each bar indicate the number of cultivars belonging to that category

Basmati 370 report values of 16.5 g (Singh et al. 2005) and 16.59 g (Sharma 2002). The values obtained in the present study are comparable with the reported values.

The germplasm assessed includes cultivars possessing superior traits in terms of physical characteristics.

3.3.2 Determination of Chemical Properties

Cooking and eating characteristics are largely determined by the properties of the starch that makes up to 90 % of the milled rice. Apparent amylose content, gelatinization temperature, and gel consistency are the important starch properties.

3.3.2.1 Determination of Apparent Amylose Content

Amylose content is the most important chemical characteristic and determines the hardness of cooked rice. Among the cultivars assessed, apparent amylose content varied from 15 % (Shrikant) to 33.19 % (Velkat) in nonbasmati scented type, 21.52 % (Basmati 370) to 26.11 % (Pusa basmati-1) in basmati type, and 22.56 % (Manila) to 31.30 % (Chitak bhat) in nonscented type (Table 3.6). Among nonbasmati scented type, state-wise, apparent amylose content varied from 17.55 % (Kala bhat) to 33.08 % (Gham) in cultivars from Maharashtra, 18.97 % (Basumati) to 29.52 % (Kagisali) in cultivars from Karnataka, and 15 % (Shrikant) to 33.19 % (Velkat) in cultivars from other Indian states. As seen from Fig. 3.6, high and intermediate amylose content was recorded in bas-

Table 3.6 Determination of chemical properties in rice (*O. sativa* L.) cultivars under study

Cultivar	Apparent amylose content ± SD (%)	Gel length (mm)	ASV
Ambemohar	22.56±0.41	39.5	4.17
Ambemohar Ajra	23.20±0.57	45.0	2.33
Ambemohar Pandhara	25.63±0.23	95.0	5.00
Ambemohar-Tambda	27.54±0.83	34.0	3.50
Champakali	25.41±1.13	36.5	2.33
Chimansal	22.75±0.23	68.0	2.17
Gham	33.08±0.53	82.0	4.50
Ghansal	26.13±0.04	83.0	2.67
Girga	28.00±1.24	95.5	4.17
Jiri	24.05±0.04	34.0	6.00
Kala bhat	17.55±0.19	64.0	3.67
Kalsal	17.69±0.30	75.5	2.50
Kamavatya	22.75±0.23	35.5	2.17
Kamod	23.57±0.19	39.0	4.67
Kate chinoor	24.24±1.13	35.5	2.50
Khadkya	26.05±3.99	48.0	3.67
Kondhekar chinoor	26.40±1.02	34.0	2.00
Kothmirsal	23.26±0.87	77.0	5.17
Lal dodki	24.24±0.00	89.5	4.33
Lal bhat	25.52±0.98	75.0	5.83
Parabhani chinoor	22.48±0.60	30.5	4.00
Raibhog	23.28±0.23	48.5	2.83
Tamsal	18.25±0.26	40.5	5.00
Velchi	23.52±1.39	81.0	2.00
Ambemohar-157	22.96±0.15	39.0	2.00
Indrayani	17.69±0.15	88.5	2.17
Ambemori	22.27±1.21	43.5	2.50
Basumati	18.97±0.75	34.0	4.00
Gandhesale	23.10±0.94	34.5	4.17
Geerige sanna	28.42±0.49	36.5	4.17
Kagisali	29.52±0.60	75.0	3.00
Kaligajvili	20.91±0.72	34.0	2.17
Kali kumud	26.77±0.11	38.5	5.17
Kumud	26.64±1.13	35.0	5.00
Medhini sanna bhatta	21.10±1.36	39.5	3.67
Mysore mallige	22.30±1.62	39.5	3.00
Mysore sanna	29.14±2.94	33.5	4.83
Sanna bili bhatta	23.49±0.68	31.5	5.17
Vasane sanna bhatta	28.10±0.11	33.5	3.67
Acharmati	26.98±0.49	36.0	3.00
Badshahbhog	20.86±0.87	38.5	4.17
Barke bhat	22.14±1.47	44.0	2.33
Bela blue	28.26±1.70	36.0	2.67
Dubraj	25.63±0.30	29.0	2.17
Dusara	23.81±0.23	71.0	7.00
Elaichi	22.00±0.60	45.0	3.50
Jeera-sona	23.26±2.00	35.0	5.00
Kalanamak	23.71±0.83	41.0	4.83
Kali kajari	24.40±1.43	69.0	7.00
Kothimbiri	21.55±2.37	80.0	4.33
Pakhe bhat	25.31±1.58	45.0	2.67
Ratibhog	17.79±1.36	61.5	3.83
Shrikamal	22.67±1.09	66.0	4.33
Shrikant	15.00±2.30	59.5	2.33
Tulshimanjula	21.10±2.11	37.5	3.67
Tulsi local	27.81±1.21	81.0	3.83
Velkat	33.19±3.16	72.0	2.00
Kolamb	26.43±1.36	94.5	3.00
Chitak bhat	31.30±0.94	78.0	2.00
Manila	22.56±1.47	39.0	4.33
Sahyadri	28.08±0.91	35.5	5.50
Pakistan basmati	22.30±0.34	44.5	6.40
Basmati 370	21.52±0.60	30.5	6.17
Pusa basmati-1	26.11±0.24	31.0	6.50
Average	24.10	51.91	3.82
CV %	15.07	39.74	36.06

mati, nonscented, and majority of nonbasmati scented rice cultivars assessed. Kala bhat, Kalsal, Tamsal, Basumati, Ratibhog, Shrikant, and Indrayani recorded low apparent amylose content.

Ramarathanam and Kulkarni (1988) have reported amylose content of 21.5 % in Ambemohar. Reports on amylose content of Badshahbhog vary between 20.3 % (Dipti et al. 2002) and 22.90 % (Shobharani and Singh 2003). These values are comparable with the values recorded in the present study. Bhatupadya et al. (2008) have noted amylose content of Medhini sanna bhatta, Mysore sanna, and Vasane Sanna Bhatta as 25.5, 25.6, and 22.8 % respectively. However, in the present analysis, Medhini sanna bhatta recorded a lower value and other cultivars recorded higher values for amylose content. This variation might be due to variation in the locality of cultivation. Earlier reports on amylose content in Basmati 370 vary from 14.7 to 22.5 % (Singh et al. 2005; Sharma 2002; Juliano and Villareal 1993), which are comparable with the present

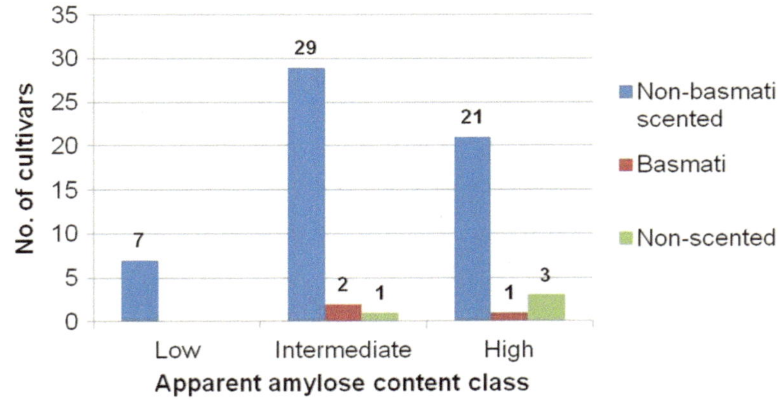

Fig. 3.6 *Classification of rice varieties under study on the basis of apparent amylose content.* The numerals over each bar indicate the number of cultivars belonging to that category

values. Rice with high amylose content is characterized by high volume expansion and cooks dry; grains are less tender and become hard on cooking. In contrast, low amylose rice cooks moist and sticky. Intermediate amylose rice has the fluffiness of high-amylose types but retains a soft texture after cooking.

In the present study, majority of the cultivars belong to intermediate amylose content. Intermediate amylose rice types are the preferred types in most of the rice-growing areas of the world and considered for the development of grain quality improvement programs.

3.3.2.2 Determination of Gel Consistency

Gel consistency is based on the consistency of the rice paste. It measures the tendency of cooked rice to harden after cooking. Selective scented rice cultivars showing varied gel consistency are presented in Fig. 3.7. Gel length of cultivars varied from 29 (Dubraj) to 95.5 mm (Girga) in nonbasmati scented rice cultivars, 30.5 (Basmati 370) to 44.5 mm (Pakistan basmati) in basmati type, and 35.5 (Sahyadri) to 94.5 mm (Kolamb) in nonscented rice cultivars (Table 3.4). Among nonbasmati scented type, gel length varied from 30.5 (Parabhani chinoor) to 99.5 mm (Girga) in cultivars from Maharashtra, 31.5 (Sanna bili bhatta) to 75 mm (Kagisali) in cultivars from Karnataka, and 29 (Dubraj) to 81 mm (Tulsi local) in cultivars from other Indian states.

As seen from Fig. 3.8, among nonbasmati scented rice type, hard, medium, and soft gel consistency was exhibited by 27, 10, and 20 cultivars respectively. Basmati cultivars recorded either high or medium gel consistency, while nonscented rice cultivars recorded hard or soft gel consistency.

On the basis of gel length, 25 high amylose rice cultivars were further classified as of hard (13 cultivars), medium (2 cultivars), and soft (10 cultivars) gel consistency (Table 3.4).

Gel length of Kalanamak is reported to vary from 25 to 38.5 mm by Singh et al. (2003). Gel length of 41 mm recorded for Kalanamak in the present study is comparable with their report. Among basmati type, gel length recorded for Basmati 370 in the present study is comparable with the value reported by Singh et al. (2000b). Sharma (2002) reported 100-mm gel length in Basmati 370, which is higher than the observed value. The gel length of Pusa basmati-1 is reported to vary from 49 (Sharma 2002) to 100 mm (Shobharani and Singh 2003), which is higher than the value recorded in the present study.

Varieties with softer gel consistency within the same amylose group are generally preferred by most rice consumers (Khush et al. 1979). As seen from Table 3.4, eight cultivars of intermediate amylose (Chimansal, Dusara, Kothmirsal, Kothimbiri, Kali kajari, Lal dodki, Shrikamal, and Velchi) and ten cultivars of high amylose (Ambemohar pandhara, Ghansal, Girga, Gham, Kagisali, Lal bhat, Tulsi local, Velkat, Chitak bhat, and Kolamb) had soft gel consistency.

3.3.2.3 Determination of Gelatinization Temperature (GT) by Alkali Digestibility Test

Gelatinization temperature (GT) determines to a large degree the suitability of rice flour as an ingredient in baked and reconstituted foods. GT

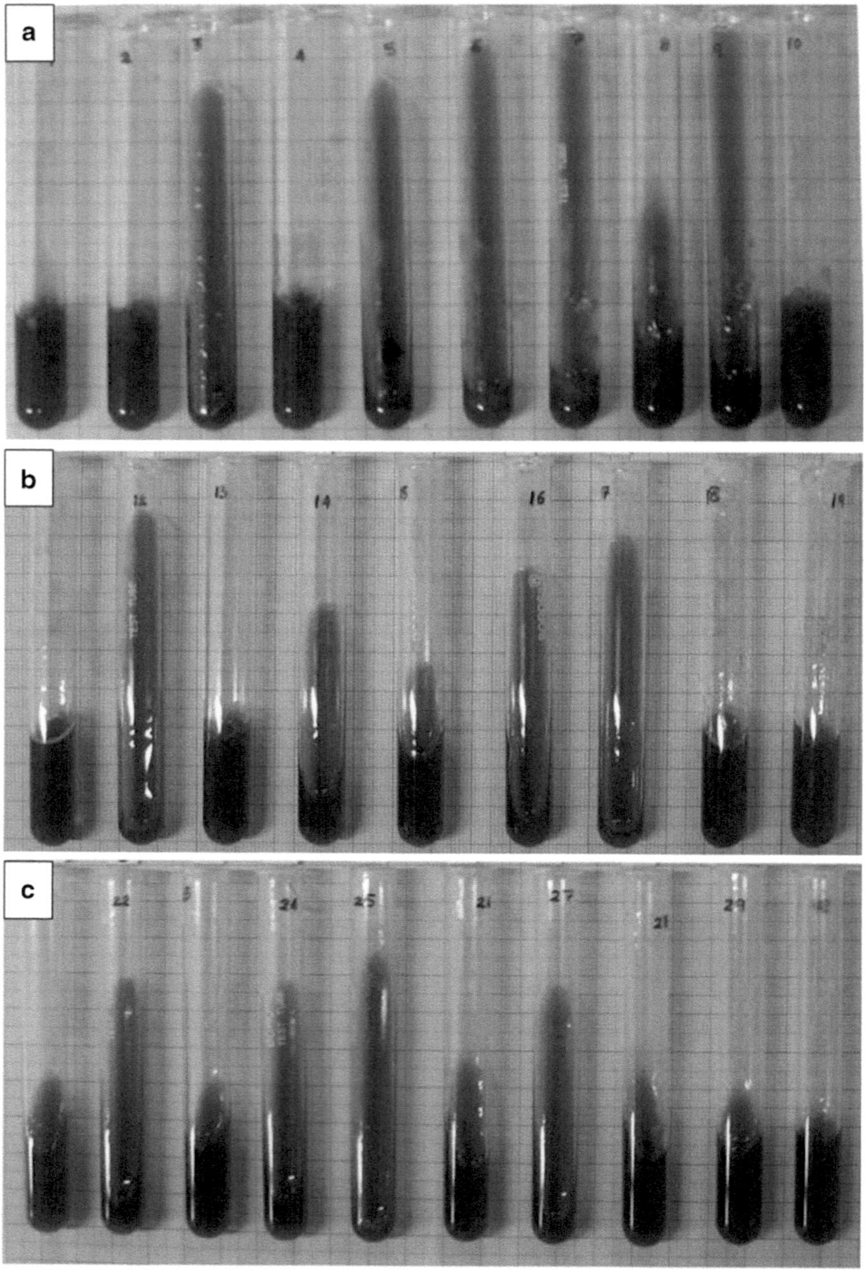

Fig. 3.7 *Selective scented rice cultivars showing varied gel consistency*. (**a**) Ambemohar, Ambemohar-Tambda, Ambemohar pandhra, Ambemohar Ajra, Dusara, Ghansal, Girga, Khadkya, Lal dodki, Acharmati (*left* to *right*). (**b**) Gandhasale, Gham, Basumati, Kala bhat, Barke bhat, Kalsal, Kothimbiri, Geerige sanna, Kamavatya (*left* to *right*). (**c**) Mysore mallige, Kali kajari, Tamsal, Shrikamal, Velkat, Pakhe bhat, Medhini sanna bhatta, Basmati 370 (*left* to *right*). (**d**) Champakali, Giri, Kothmirsal, Raibhog, Makarand, Laal Bhat, Kumud, Umrani, Manila, Chittak bhat (*left* to *right*). (**e**). Ambemohar-157, Basmati, Indrayani, Sahyadri, Pusa basmati-1, Kolamb, Sonamasuri (*left* to *right*)

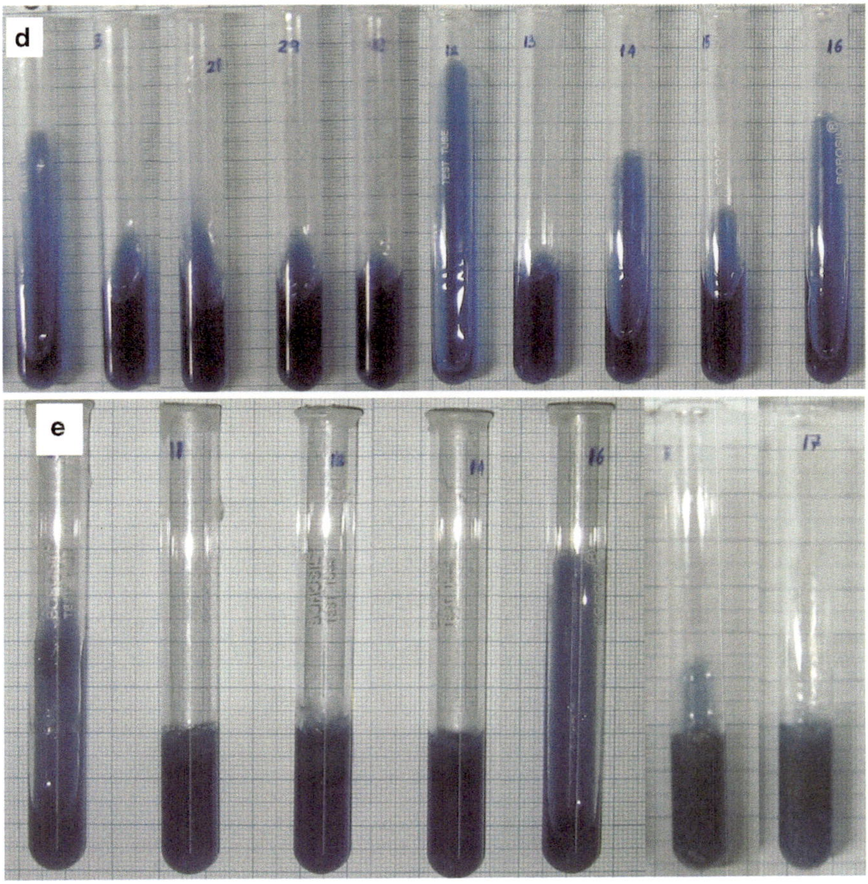

Fig. 3.7 (continued)

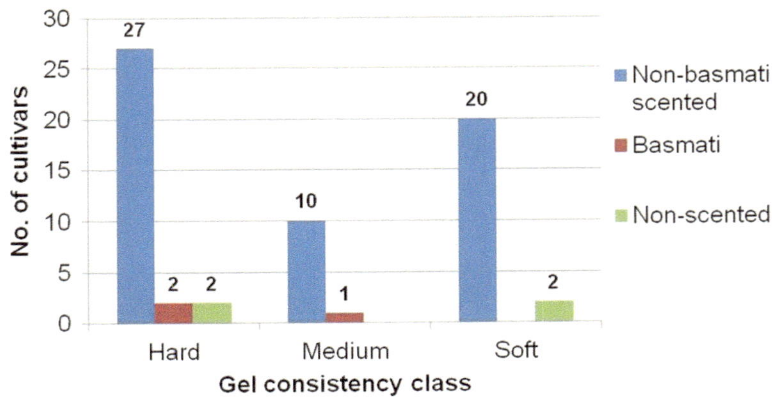

Fig. 3.8 *Classification of rice varieties under study on the basis of gel consistency*. The numerals over each bar indicate the number of cultivars belonging to that category

is inversely proportional to alkali spreading value (ASV). Selective scented rice cultivars showing varied ASV are presented in Fig. 3.9.

Among cultivars, ASV varied from 2 (Velchi, Ambemohar-157, Velkat) to 7 (Dusara, Kali kajari) in nonbasmati scented rice cultivars, 6.17 (Basmati 370) to 6.5 (Pusa basmati-1) in basmati, and 2 (Chitak bhat) to 5.5 (Sahyadri) in nonscented rice cultivars (Table 3.4). Among nonbasmati scented type, ASV varied from 2 (Velchi, Ambemohar-157) to 6 (Jiri), 2.17 (Kaligajvili) to 5.17 (Kali kumud), and 2 (Velkat) to 7 (Dusara, Kali kajari) in cultivars from Maharashtra, Karnataka, and other Indian states respectively. As seen from Fig. 3.10, majority of the nonbasmati scented rice cultivars belong either to intermediate (30 cultivars) or high

a

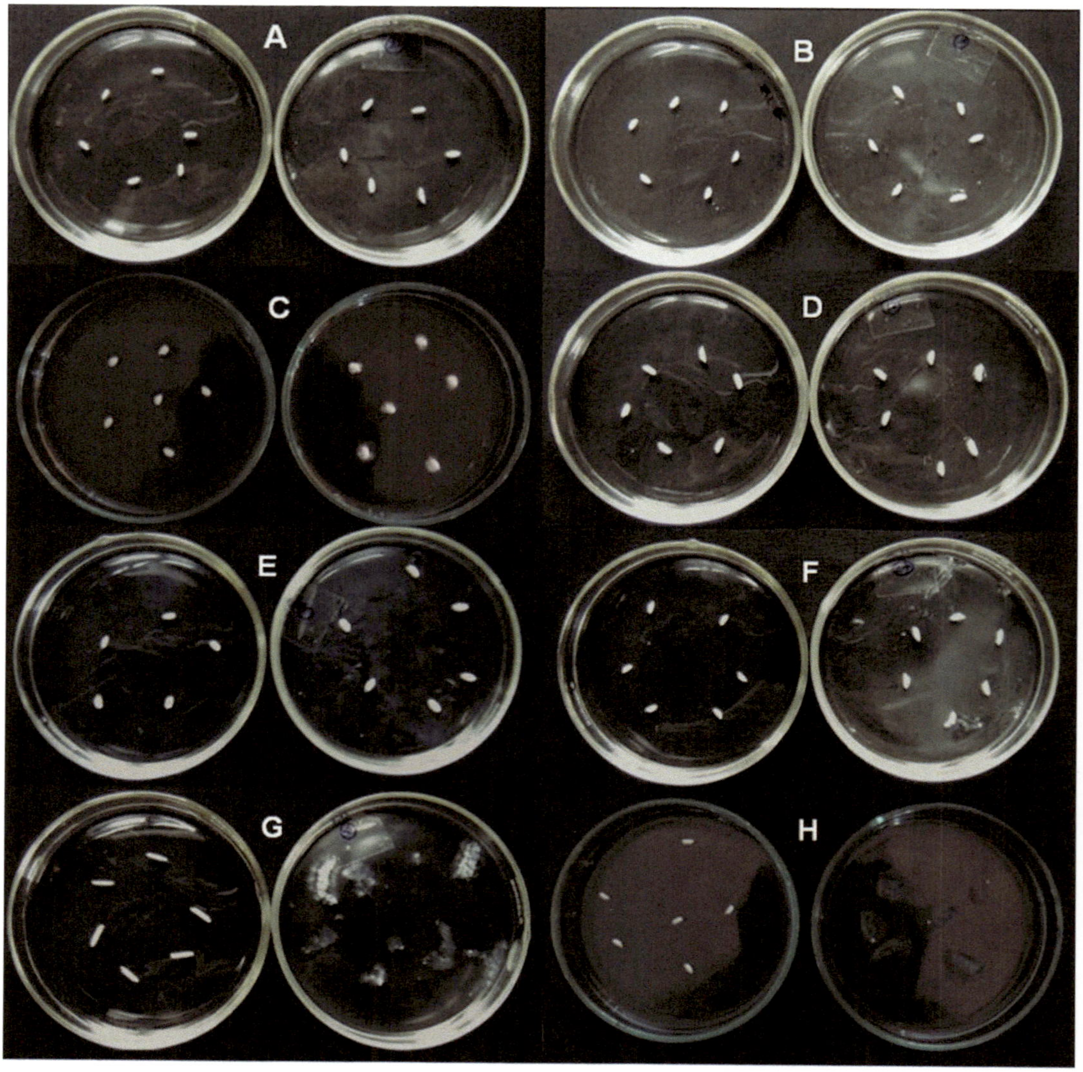

Fig. 3.9 *Selective scented rice cultivars showing varied alkali spreading values.* (**a**) *A*: Ambemohar Ajra, *B*: Ghansal, *C*: Acharmati, *D*: Khadkya, *E*: Girga, *F*: Kumud, *G*: Basmati 370, *H*: Dusara (*left side* Petri plate represents control), (**b**) *A*: Tulsi Local, *B*: Tamsali, *C*: Jiri, *D*: Basmati, *E*: Ambemohar-Tambda, *F*: Ambemohar 157, *G*: Ambemohar, *H*: Champakali, (**c**) *A*: Elichi, *B*: Kothimbiri, *C*: Ratibhog, *D*: Shrikamal, *E*: Medhini Sanna Bhatta, *F*: Bella Blue, *G*: Tulshimanjula, *H*: Kali Kajari, (**d**) *A*: Indrayani, *B*: Kothmirsal, *C*: Sonamasuri, *D*: Pusa basmati-1, *E*: Makarand, *F*: Laal Bhat, *G*: Jiri, *H*: Gham

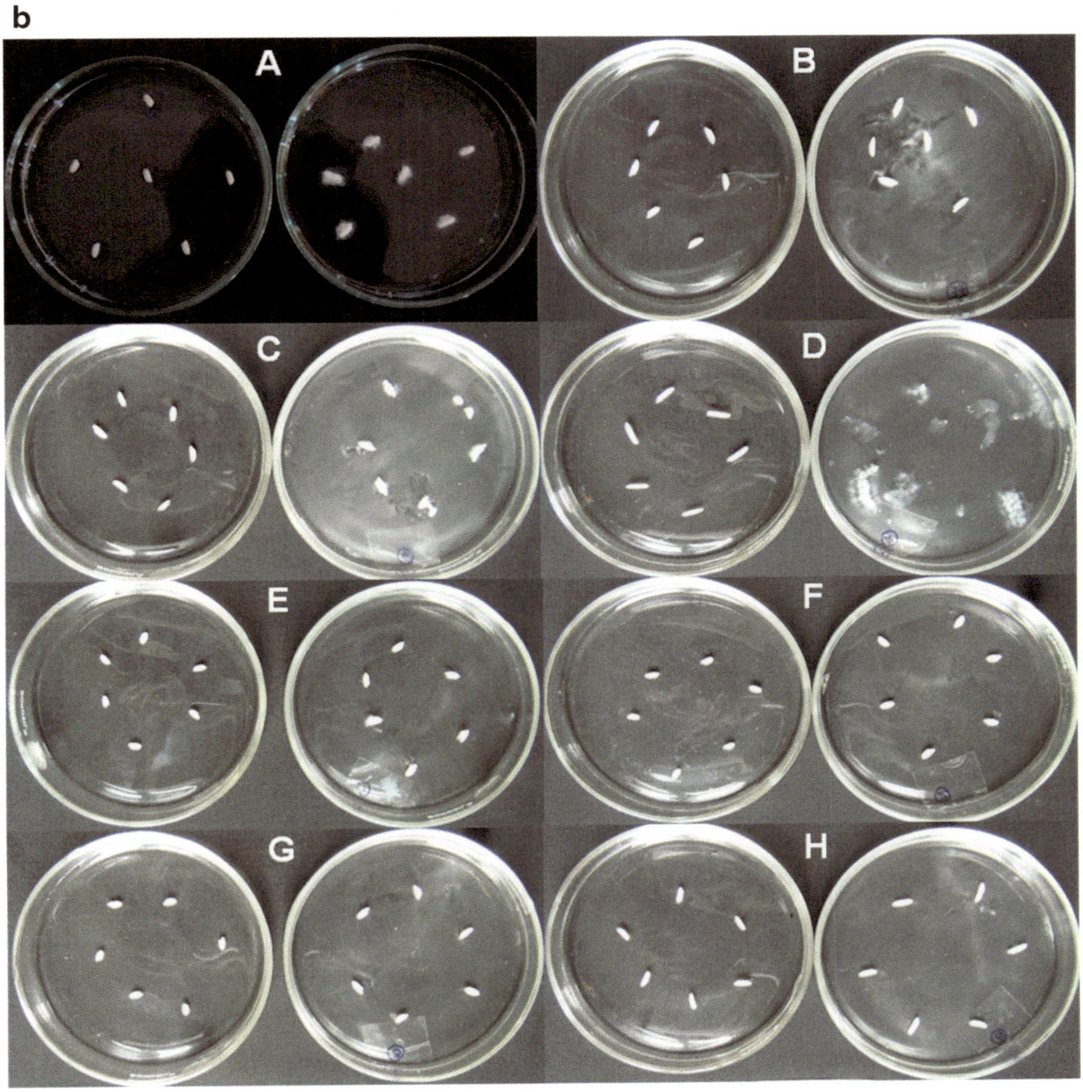

Fig. 3.9 (continued)

c

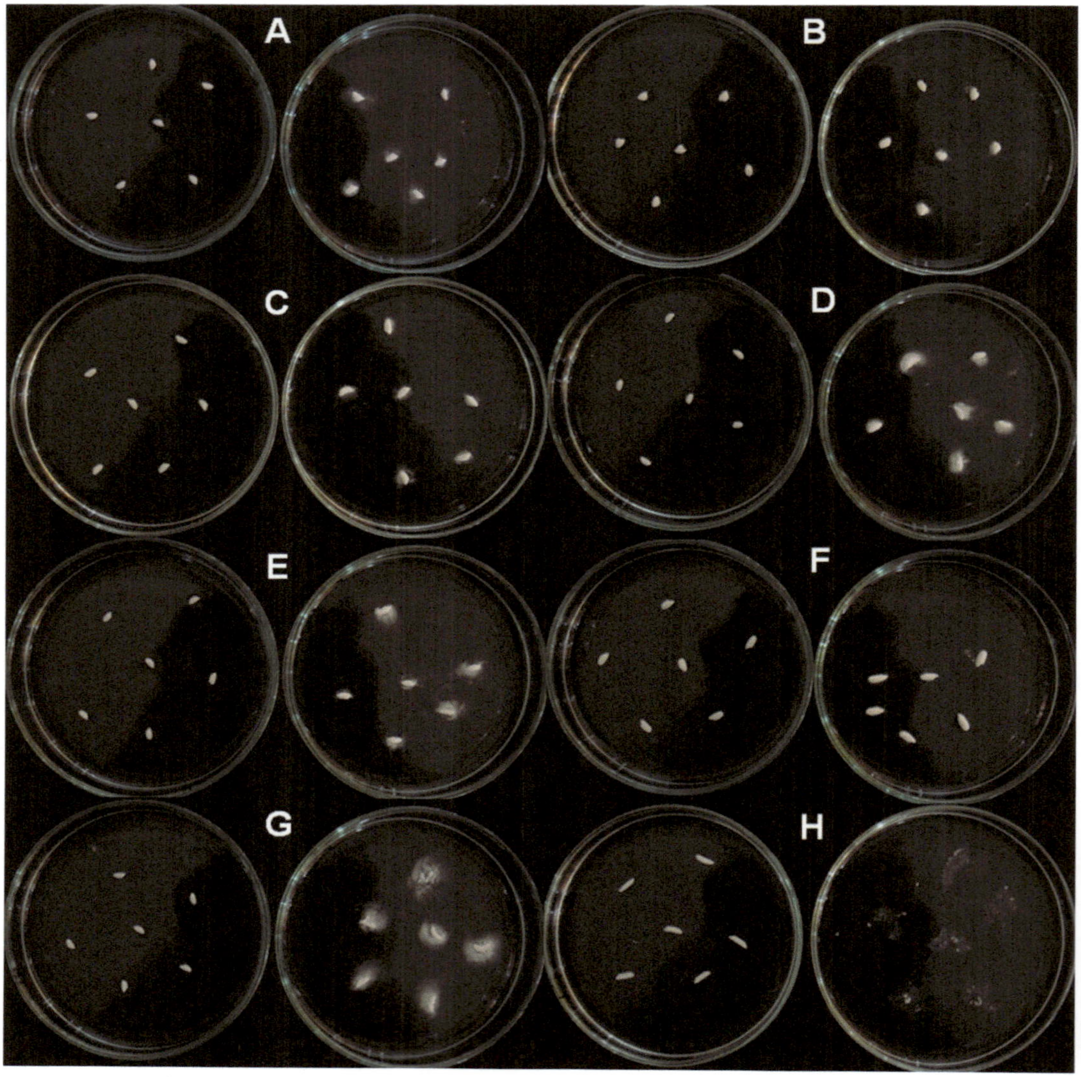

Fig. 3.9 (continued)

Fig. 3.9 (continued)

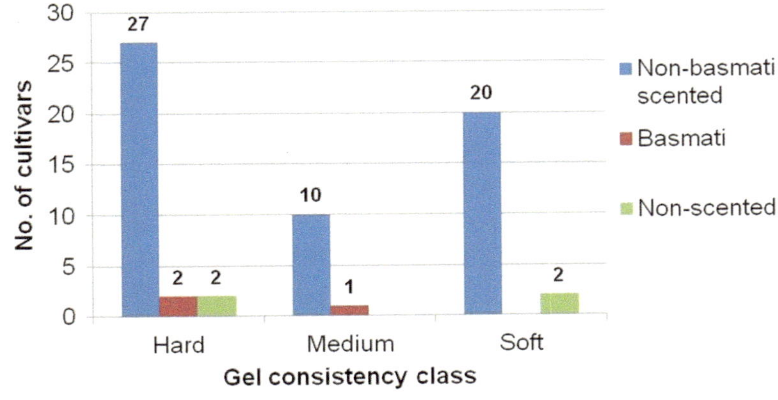

Fig. 3.10 *Classification of rice varieties under study on the basis of gelatinization temperature*. The numerals over each bar indicate the number of cultivars belonging to that category

(16 cultivars) GT class. Basmati type exhibited low GT, while each of the nonscented rice cultivars was categorized into different GT class. Among nonbasmati scented rice cultivars, Jiri, Lal bhat, Dusara and Kali kajari recorded low GT.

In the present study, ASV of 4.17 is recorded in Badshahbhog. It is comparable with the reported values of 3.6 (Dipti et al. 2002) and 4–6.33 (Bisne and Sarawgi 2008) in Badshahbhog. ASV of Kalanamak and Dubraj is reported to vary from 3.7 to 4 (Singh et al. 2000a) and 2.5 to 2.7 (Marothia 2003).

Several reports are available on ASV of Basmati 370. These reports revealed that ASV of Basmati 370 varies: 3.5 (Sharma 2002), 3.7 (Singh et al. 2000b), 4.5 (Shobharani et al. 2006), and 6.2 (Shobharani and Singh 2003). ASV of 7 is reported in Pusa basmati-1 by Sharma (2002) and Shobharani et al. (2006). Present study revealed comparable values for ASV in these cultivars.

Many food processing applications prefer rice with intermediate gelatinization temperature (Mutters and Thompsons 2009). Cooking and eating qualities can be improved by incorporating intermediate amylose content, intermediate starch gelatinization temperature, and soft gel consistency (Khush and Juliano 1985). Kothmirsal, Kothimbiri, Lal dodki, and Shrikamal possess these qualities.

3.3.3 Determination of Cooking Characteristics

Rice is consumed as a cooked whole grain by majority of rice consumers. Cooking and eating characteristics of rice play an important role consumer preference. Selective scented rice cultivars showing longitudinal and latitudinal expansion after cooking are presented in Fig. 3.11.

As depicted in Table 3.7, maximum length of cooked grain was reported by Raibhog (11.85 mm), followed by Sahyadri (11.55 mm) and Gham (11.36 mm). These values are higher than that of Basmati 370 (10.93 mm). Kali kumud recorded lowest value of 6.55 mm for cooked rice length. Kalanamak and Ambemohar Ajra exhibited lower values for breadth of cooked grain (1.93 and 2.36 mm respectively), which are comparable with Basmati 370 (2.03 mm) and Pusa basmati-1 (1.93 mm).

Cooked grain with higher length and lower breadth are preferred by majority of Indian consumers and receive higher market demand. Cooking coefficient indicates longitudinal as well as latitudinal variations in the grain dimensions from raw rice to cooked rice. Cooking coefficient of the nonbasmati scented cultivars varied from 1.27 (Tulsi local) to 4.98 (Kothimbiri). Basmati 370 recorded cooking coefficient of 1.73. Resultant variation in the grain length after cooking was determined by coefficient of elongation. Among nonbasmati cultivars, coefficient of elongation recorded variation from 1.25 (Tulsi local) to 2.45 (Raibhog). Kothimbiri, along with Ambemohar-157, Ambemohar-Tambda, and Ambemohar Ajra, recorded higher elongation as seen from Table 3.5. Medhini sanna bhatta and Parabhani chinoor recorded lowest (1.03) and highest (1.63) values, respectively, for latitudinal expansion.

Basmati rice is preferred over nonbasmati rice due to higher elongation and lower latitudinal expansion. In the present study, Basmati 370 had elongation coefficient of 1.55 and latitudinal expansion of 1.07. Earlier reports on elongation of Basmati 370 vary as 1.00 (Richharia and Govindswami 1966), 1.42 (Singh et al. 2005), 1.7 (Sharma 2002), 1.85 (Shobharani et al. 2006), and 1.94 (Singh 2000). Elongation is influenced by genetic factors, and environment (Dela Cruz et al. 1989) explains the variation in value of coefficient of elongation. Elongation in Badshahbhog (Table 3.2) is comparable with the value of 1.5 reported by Dipti et al. (2002) and lower than the value reported by Malik et al. (1994). Cooked rice is a final product which is consumed, irrespective of dimensions of uncooked rice, so L/B ratio of cooked grain was determined. Basmati 370 recorded highest cooked L/B ratio of 5.38, followed by Pusa basmati-1 (4.3). Kate chinoor (4.25), Kondhekar chinoor (4.25), and Basumati (4.23) also recorded higher values for this trait. Minimum cooked L/B ratio was observed in Acharmati (1.9) and Mysore sanna (2.12). Cooked L/B ratio of Basmati 370 is reported as 4.33 by Singh et al. (2005), which is lower than the present study.

A comparative account of cooking characteristics among nonbasmati scented, basmati, and nonscented rice cultivars is presented as Table 3.6

As seen from Table 3.8, nonbasmati scented cultivars excel than basmati in some of the cooking characteristics. Basmati cultivars were selected and improved by local farmers for their superior quality characteristics over other cultivars (Siddiq et al. 2012). It was observed from results that traditional landraces, viz., Raibhog, Kate chinoor, Kondhekar chinoor, Basumati, Kothimbiri, Ambemohar Ajra, Ambemohar-Tambda, and Gham, possess superior cooking

Fig. 3.11 *Selective scented rice cultivars showing longitudinal and latitudinal expansion after cooking*. Unhusked grains (*upper row*), dehusked uncooked grains (*middle row*), and cooked grains (*lower row*) of (**a**) *A*: Ambemohar Ajra, *B*: Kothimbiri, *C*: Raibhog, *D*: Ambemohar-Tambda, *E*: Basumati, *F*: Kate chinoor, *G*: Gham, *H*: Basmati 370, *I*: Ghansal, *J*: Chimansal, *K*: Parabhani chinoor, *L*: Kalsal, *M*: Dubraj, *N*: Kalanamak, (**b**) *A*: Ambemohar, *B*: Ambemohar pandhra, *C*: Acharmati, *D*: Chinikamini, *E*: Elichi, *F*: Tulsiamrit, *G*: Girija Sambha, *H*: Tulsimanjula, *I*: Dusra, *J*: Shrikant, *K*: Shrikamal, *L*: Gopalbhog, *M*: Giri, *N*: Girja, *O*: Girikolpi, *P*: Kali Kumud, *Q*: Gandhasale, *R*: Mysore Sanna, (**c**) *A*: Kondharkar Chinnor, *B*: Kolamb, *C*: Pusa Sughandha, *D*: Kamod, *E*: Lal Dodki, *F*: Indrayani, *G*: Jeerasona, *H*: Gerrige Sanna, *I*: Kalabhat, *J*: Khadkya, *K*: Kamavtya, *L*: Kolamb, *M*: Tamsal

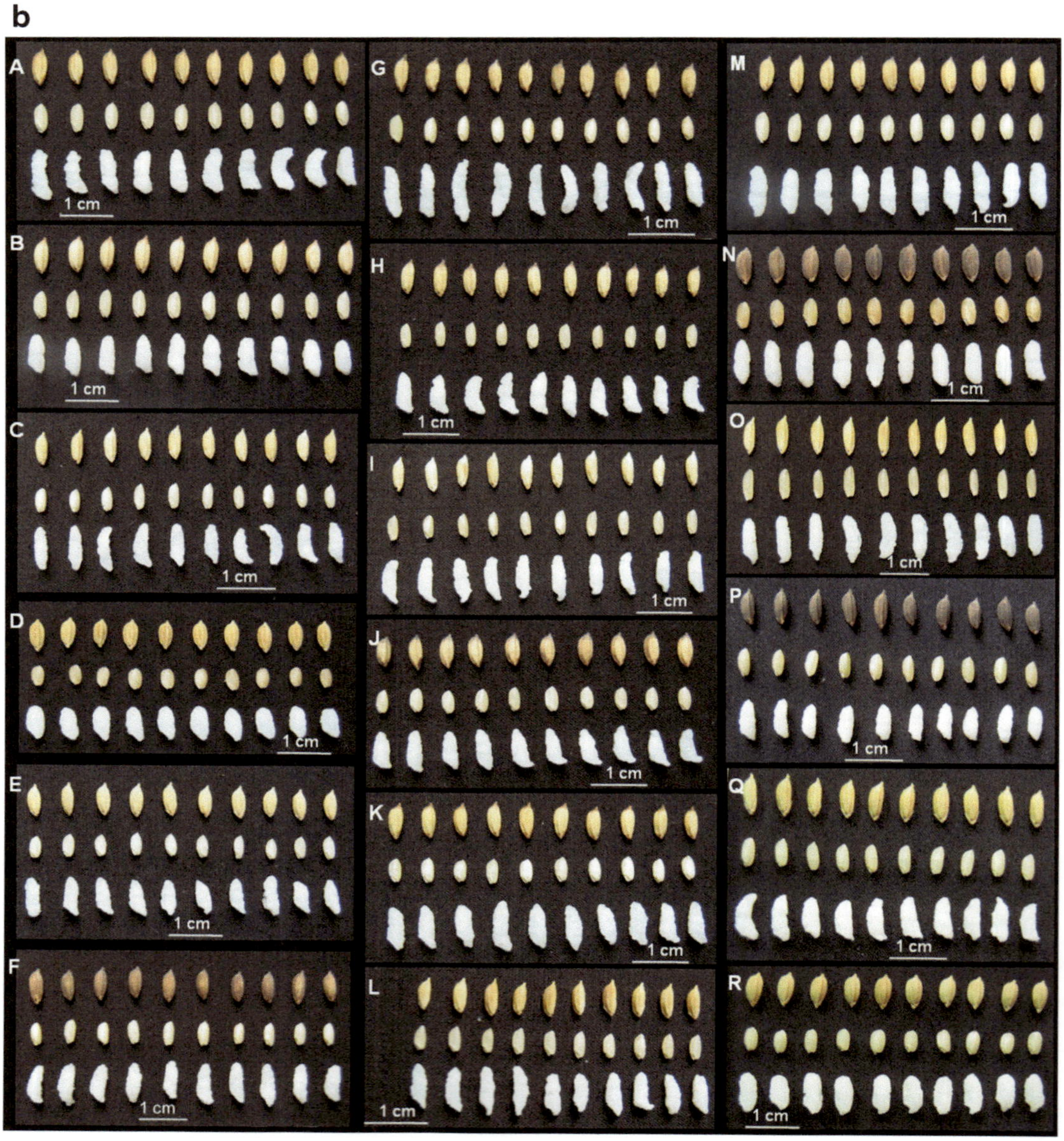

Fig. 3.11 (continued)

c

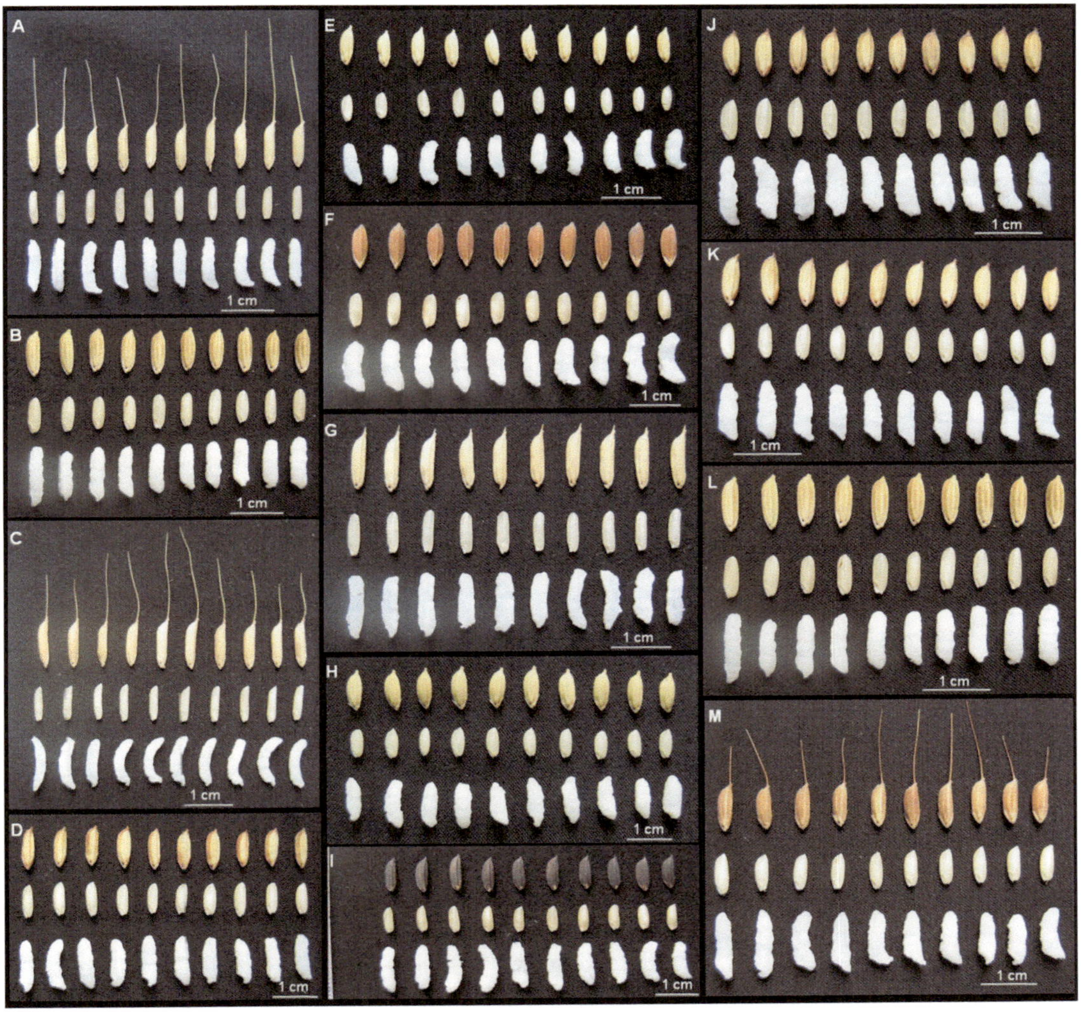

Fig. 3.11 (continued)

Table 3.7 Determination of cooking characteristics in rice (*O. sativa* L.) varieties under study

Cultivar	Av. Lc ± SD (mm)	Av. Bc ± SD (mm)	CC	CE	CLF	Lc/Bc
Ambemohar	8.11 ± 0.44	2.97 ± 0.09	2.18	1.71	1.25	2.73
Ambemohar Ajra	7.40 ± 0.47	2.36 ± 0.10	2.18	1.81	1.33	3.14
Ambemohar Pandhara	7.43 ± 0.31	3.05 ± 0.08	1.76	1.55	1.33	2.44
Ambemohar-Tambda	9.01 ± 0.31	3.63 ± 0.20	2.33	1.81	1.36	2.48
Champakali	9.55 ± 0.41	3.15 ± 0.17	1.58	1.58	1.58	3.03
Chimansal	9.65 ± 0.46	3.07 ± 0.15	1.70	1.62	1.48	3.14
Gham	11.36 ± 0.42	3.10 ± 0.14	1.70	1.62	1.42	3.66
Ghansal	7.25 ± 0.30	2.75 ± 0.10	1.79	1.57	1.31	2.64
Girga	6.90 ± 0.37	2.81 ± 0.10	1.74	1.48	1.22	2.46
Jiri	8.33 ± 0.40	2.90 ± 0.12	1.92	1.70	1.40	2.87
Kala bhat	10.23 ± 0.47	2.71 ± 0.14	1.91	1.72	1.34	3.77
Kalsal	9.33 ± 0.50	2.67 ± 0.08	1.72	1.51	1.15	3.49
Kamavatya	9.13 ± 0.28	2.68 ± 0.15	1.81	1.63	1.32	3.41
Kamod	10.35 ± 0.49	2.91 ± 0.12	2.03	1.72	1.24	3.56
Kate chinoor	10.76 ± 0.51	2.53 ± 0.05	1.70	1.67	1.58	4.25
Khadkya	8.92 ± 0.44	2.83 ± 0.18	1.63	1.52	1.33	3.15
Kondhekar chinoor	10.88 ± 0.66	2.56 ± 0.05	1.80	1.73	1.52	4.25
Kothmirsal	7.48 ± 0.28	2.65 ± 0.14	2.15	1.74	1.29	2.82
Lal dodki	10.49 ± 0.59	3.55 ± 0.05	1.94	1.68	1.34	2.95
Lal bhat	10.55 ± 0.38	3.90 ± 0.34	1.78	1.62	1.41	2.71
Parabhani chinoor	10.38 ± 0.50	2.59 ± 0.09	1.74	1.71	1.63	4.01
Raibhog	11.85 ± 0.44	3.42 ± 0.06	3.27	2.45	1.51	3.46
Tamsal	10.14 ± 1.14	2.79 ± 0.12	1.92	1.67	1.24	3.63
Velchi	8.42 ± 0.20	2.62 ± 0.08	1.70	1.50	1.18	3.21
Ambemohar-157	9.28 ± 0.29	3.42 ± 0.08	2.38	1.88	1.39	2.71
Indrayani	11.10 ± 0.73	2.98 ± 0.16	1.66	1.61	1.49	3.72
Ambemori	6.74 ± 0.44	2.79 ± 0.18	1.98	1.54	1.17	2.42
Basumati	10.74 ± 0.67	2.54 ± 0.07	1.77	1.60	1.23	4.23
Gandhesale	6.90 ± 0.37	2.75 ± 0.20	1.65	1.48	1.27	2.51
Geerige sanna	6.75 ± 0.49	2.55 ± 0.21	1.80	1.53	1.23	2.65
Kagisali	7.71 ± 0.47	2.94 ± 0.30	1.33	1.26	1.16	2.62
Kaligajvili	9.98 ± 0.74	2.87 ± 0.13	1.54	1.45	1.28	3.48
Kali kumud	6.55 ± 0.35	2.57 ± 0.09	1.60	1.44	1.24	2.55
Kumud	7.50 ± 0.25	2.68 ± 0.12	1.96	1.60	1.21	2.80
Medhini sanna bhatta	6.97 ± 0.24	2.49 ± 0.07	1.88	1.45	1.03	2.80
Mysore mallige	8.91 ± 0.59	2.93 ± 0.21	1.58	1.45	1.25	3.04
Mysore sanna	7.45 ± 0.28	3.51 ± 0.10	2.17	1.63	1.27	2.12
Sanna bili bhatta	9.21 ± 0.40	3.40 ± 0.23	1.29	1.28	1.28	2.71
Vasane sanna bhatta	7.07 ± 0.59	2.80 ± 0.35	2.68	1.71	1.10	2.53
Acharmati	6.86 ± 0.14	3.62 ± 0.09	1.99	1.68	1.47	1.90
Badshahbhog	6.65 ± 0.33	2.65 ± 0.22	1.86	1.55	1.24	2.51
Barke bhat	8.56 ± 0.42	3.19 ± 0.20	1.55	1.42	1.24	2.68
Bela blue	8.37 ± 0.38	3.40 ± 0.14	1.71	1.43	1.16	2.46
Dubraj	8.81 ± 0.17	3.00 ± 0.11	1.50	1.50	1.51	2.94
Dusara	7.40 ± 0.09	2.46 ± 0.07	1.87	1.64	1.33	3.01
Elaichi	7.07 ± 0.18	3.01 ± 0.17	1.95	1.67	1.40	2.35

(continued)

Table 3.7 (continued)

Cultivar	Av. Lc±SD (mm)	Av. Bc±SD (mm)	CC	CE	CLF	Lc/Bc
Jeera-sona	6.82±0.13	2.85±0.14	1.58	1.54	1.48	2.39
Kalanamak	8.11±0.19	1.93±0.09	1.72	1.50	1.06	4.20
Kali kajari	9.16±0.30	2.80±0.23	1.32	1.34	1.39	3.27
Kothimbiri	8.53±0.34	3.00±0.14	4.98	2.19	1.08	2.84
Pakhe bhat	9.83±0.12	3.24±0.11	1.56	1.50	1.41	3.03
Ratibhog	7.36±0.34	2.90±0.21	1.73	1.52	1.28	2.54
Shrikamal	7.41±0.10	3.24±0.05	2.04	1.74	1.47	2.29
Shrikant	7.19±0.12	2.63±0.16	2.17	1.72	1.27	2.73
Tulshimanjula	7.01±0.15	2.90±0.14	1.73	1.59	1.42	2.42
Tulsi local	6.85±0.2	2.84±0.25	1.27	1.25	1.22	2.41
Velkat	7.90±0.29	3.00±0.22	1.46	1.39	1.27	2.63
Kolamb	9.53±0.39	2.71±0.09	1.77	1.55	1.18	3.52
Chitak bhat	11.07±0.52	3.95±0.17	1.79	1.63	1.40	2.80
Manila	10.39±0.40	3.38±0.15	1.64	1.55	1.38	3.07
Sahyadri	11.55±0.44	3.25±0.26	1.63	1.56	1.40	3.55
Pakistan basmati	8.39±0.29	3.52±0.16	1.27	1.38	1.58	2.38
Basmati 370	10.93±0.38	2.03±0.11	1.73	1.55	1.07	5.38
Pusa basmati-1	8.30±0.29	1.93±0.12	1.02	1.02	1.01	4.30

Av. Average, *Lc* Length of cooked grain, *Bc* Breadth of cooked grain, *CC* Coefficient of cooking, *CE* Coefficient of elongation, *CLE* Coefficient of latitudinal expansion

Table 3.8 Comparative account of cooking characteristics of basmati and nonbasmati scented rice (*O. sativa* L.) cultivars

Parameter	Nonbasmati scented	Basmati	Nonscented
Length of cooked kernel (mm)	6.55 (Kali kumud) – 11.85 (Raibhog)	8.30 (Pusa basmati-1) – 10.93 (Basmati 370)	9.53 (Kolamb) – 11.55 (Sahyadri)
Breadth of cooked kernel (mm)	1.93 (Kalanamak) – 3.90 (Lal bhat)	1.93 (Pusa basmati-1) – 3.52 (Pakistan basmati)	2.71 (Kolamb) – 3.95 (Chitak bhat)
Coefficient of cooking	1.27 (Tulsi local) – 4.98 (Kothimbiri)	1.02 (Pusa basmati-1) – 1.73 (Basmati 370)	1.63 (Sahyadri) – 1.79 (Chitak bhat)
Coefficient of elongation	2.39 (Acharmati) – 27.19 (Medhini sanna bhatta)	1.80 (Pusa basmati-1) – 30.00 (Basmati 370)	3.77 (Chitak bhat) – 8.06 (Kolamb)
Coefficient of latitudinal expansion	1.03 (Medhini sanna bhatta) – 1.63 (Parabhani chinoor)	1.01 (Pusa basmati-1) – 1.58 (Pakistan basmati)	1.18 (Kolamb) – 1.40 (Chitak bhat)

characteristics and can significantly contribute to the gene pool of scented rice cultivars.

3.3.4 Correlation between Physical, Chemical, and Cooking Characteristics

The two-tailed correlation coefficients between length (Lu) and breadth (Bu) of uncooked grain, L/B ratio (L/B), test weight (TW), length (Lc) and breadth (Bc) of cooked grain, coefficient of cooking (CC), coefficient of elongation (CE), and coefficient of latitudinal expansion (CLE), L/B ratio of cooked grain (Lc/Bc), apparent amylose content (AC), gel length (GL), and alkali spreading value (ASV) are depicted in Table 3.9. Grain length (Lu) has significant positive association with L/B ratio, test weight, length of cooked grain, and L/B ratio of cooked grain. These findings were also recorded by Khatun et al. (2003). Krishna Veni and Shobha Rani (2006) observed similar relation; however, they reported negative correlation between grain length and test weight. Grain length is negatively correlated with coefficient of cooking and coefficient of elongation.

Table 3.9 Correlation among physicochemical and cooking characteristics studied in 64 rice cultivars

Parameter	Lu	Bu	L/B	TW	Lc	Bc	CC	CE	CLE	Lc/Bc	AC	GL
Bu	−0.029											
L/B	0.808[a]	−0.591[a]										
TW	0.700[a]	0.572[a]	0.226									
Lc	0.749[a]	−0.027	0.628[a]	0.614[a]								
Bc	0.051	0.713[a]	−0.372[a]	0.534[a]	0.233							
CC	−0.503[a]	0.267[b]	−0.492[a]	−0.141	0.009	0.126						
CE	−0.421[a]	0.028	−0.315[b]	−0.166	0.266[b]	0.224	0.830[a]					
CLE	0.088	−0.301[b]	0.244	−0.014	0.349[a]	0.446[a]	−0.123	0.291[b]				
Lc/Bc	0.629[a]	−0.526[a]	0.831[a]	0.154	0.684[a]	−0.528[a]	−0.102	0.034	−0.054			
AC	0.086	0.270[b]	−0.067	0.162	−0.046	0.270[b]	−0.139	−0.195	0.022	−0.216		
GL	−0.007	0.187	−0.141	0.245	0.016	0.132	0.081	0.019	−0.066	−0.126	0.107	
ASV	0.063	−0.070	0.081	0.000	−0.094	−0.140	−0.077	−0.176	−0.143	0.066	0.029	−0.058

Lu length of uncooked grain, *Bu* breadth of uncooked grain, *L/B* L/B ratio of uncooked grain, *TW* test weight, *Lc* length of cooked grain, *Bc* breadth of cooked grain, *CC* coefficient of cooking, *CE* coefficient of elongation, *CLE* coefficient of latitudinal expansion, *Lc/Bc* Lc/Bc ratio, coefficient, *AC* apparent amylose content, *GL* gel length, *ASV* alkali spreading value

[a]Correlation is significant at the 0.01 level (two tailed)

[b]Correlation is significant at the 0.05 level (two tailed)

This relation was nonsignificant as reported by the earlier studies (Khatun et al. 2003; Krishna Veni and Shobha Rani 2006). Grain breadth correlates negatively with L/B ratio, coefficient of latitudinal expansion, and L/B of cooked grain. Existing literature (Hussain et al 1987; Khatun et al. 2003; Krishna Veni and Shobha Rani 2006) supports this finding. L/B ratio had significant positive association with length of cooked grain and L/B of cooked grain. L/B ratio correlated negatively with breadth of cooked grain, coefficient of cooking, and coefficient of elongation. In a study on aromatic rice, Khatun et al. (2003) also observed positive association between L/B ratio and length of cooked grain. Test weight correlates positively with length and breadth of cooked grain. Length of cooked grain has positive correlation with coefficient of elongation, coefficient of latitudinal expansion, and L/B ratio of cooked grain. Krishna Veni and Shobha Rani (2006) recorded positive relation between length of cooked grain and coefficient of elongation; however, Khatun et al. (2003) recorded nonsignificant relation between these variables. Breadth of cooked grain correlates positively with coefficient of latitudinal expansion and correlates negatively with L/B of cooked grain. Coefficient of elongation positively associates with coefficient of cooking and coefficient of latitudinal expansion.

Amylose content correlates positively with breadth of uncooked and cooked grains. The present study revealed that amylose content, gel length, and alkali spreading value are not associated with each other. Similar finding was also recorded by Krishna Veni and Shobha Rani (2006). This could be attributed by the diverse genetic makeup of the scented landraces assessed in the present study.

References

Adair CR, Brachell HM, Jodon NE, Johnston TH, Thysell JR, Green VE, Webb BD, Atkins JG (1966) Rice breeding and testing methods in the U.S. USDA Agricultural Research Service Handbook, 289, Washington, D.C. (USA) pp 19–64

Bhatupadya VG, Bhat RS, Shenoy VV, Salimath PM (2008) Physico-chemical characterization of popping-special rice accessions. Karnataka J Agric Sci 21(2):84–186

Bisne R, Sarawgi AK (2008) Agro-morphological and quality characterization of Badshahbhog group from aromatic rice germplasm of Chhattisgarh. Bangladesh J Agric Res 33(3):479–492

Cagampang GB, Parez CM, Juliano BO (1973) A gel consistency test for eating quality of rice. J Sci Food Agric 24:1589–1594

Cruz DN, Khush GS (2004) Rice grain quality evaluation procedures. In: Singh RK, Singh US, Khush GS (eds) Aromatic rices. Kalyani, New Delhi, pp 15–28

Dela Cruz N, Kumar I, Kaushik RP, Khush GS (1989) Effect of temperature during grain development on the performance and stability of cooking quality components of rice. Jpn J Breed 39:299–306

Dipti SS, Hossain ST, Bari MN, Kabir KA (2002) Physicochemical and cooking properties of some fine rice varieties. Pak J Nutr 1(4):188–190

Dipti SS, Bari MN, Kabir KA (2003) Grain quality characteristics of some Beruin Rice varieties of Bangladesh. Pak J Nutr 2(4):242–245

Dutta RK, Lahiri BP, Baset MA (1998) Characterization of some aromatic and fine rice cultivars in relation to their physicochemical quality of grains. Ind J Plant Phys 3(1):61–64

George SP, Bastin D, Radhakrishnan NV, Aipe KC (2005) Evaluation of aromatic rice varieties in Wayanad, Kerala. J Trop Agric 43(1–2):67–69

Hussain AA, Maurya DM, Vaish CP (1987) Studies on quality status of indigenous upland rice (*Oryza sativa*). Ind J Genet 47(2):145–152

Juliano BO (1971) A simplified assay for milled rice amylose. Cereal Sci Today 16:334–338

Juliano BO (1979) The chemical basis of rice grain quality. In: Proceedings of the workshop on chemical aspects of rice grain quality. IRRI, Los Banos, Philippines

Juliano BO, Villareal CP (1993) Grain quality evaluation of world rices. IRRI, Manila, Philippines, p 205

Juliano BO, Bautista GM, Lugay JC, Reyes AC (1964) Studies on the physicochemical properties of rice. J Agric Food Chem 12:131–138

Khatun MM, Ali MH, Delacruz QD (2003) Correlation studies on grain Physicochemical Characteristics of Aromatic Rice. Pak J Biol Sci 6(5):511–513

Khush GS, Juliano BO (1985) Rice grain quality and marketing. In: Breeding for high-yielding rices of exceilent cooking and eating quality. IRRI, Manila, pp 61–69

Khush GS, Paule CM, Delacruz NM (1979) Rice grain quality evaluation and improvement at IRRI. In: Chemical aspects of rice grain quality. IRRI, Los Baños, pp 21–31

Krishna Veni B, Shobha Rani N (2006) Association of grain yield with quality characteristics and other yield components in rice. Oryza 43(4):320–322

Kumar I, Khush GS (1986) Gene dosage effect of amylose content in rice endosperm. Jpn J Genet 61:559–568

Kumar D, Singh G, Singh US (2003) Evaluation of physical characteristics of aromatic rice grown at different locations. J Food Sci Technol 40(6):660–662

Little RR, Hilder GB, Dawson EH (1958) Differential effect of dilute alkali on 25 varieties of milled white rice. Cereal Chem 35:111–126

Mahadevappa M, Mohankumar HD (2003) The aromatic rices of Karnataka. In: Singh RK, Singh US (eds) A treatise on the scented rices of India. Kalyani publishers, New Delhi, pp 355–375

Mahindru SK (1995) Manual of basmati rice. Metropolitan Book, New Delhi, pp 16–21

Malik SS, Dikshit N, Dash AB, Lodh SB (1994) Studies on agromorphological and physicochemical characteristica of local scented rice varieties. Oryza 31:106–110

Marothia DM (2003) Domestic trade in aromatic rice: lessons from an exploratory study of Chhattisgarh. In: Singh RK, Singh US (eds) A treatise on the scented rices of India. Kalyani Publishers, New Delhi, pp 143–164

Mutters RG, Thompsons JF (2009). Rice quality in global market. In: Rice quality handbook, University of California, Agriculture and Natural Resources, Canada, p 13

Ramarathanam N, Kulkarni PR (1988) Chemical composition and cooking quality of rice varieties grown in Maharashtra. J Maharashtra Agric Univ 13(2):203–205

Richharia RH, Govindswami S (1966) Rices in India. Scientific Book Co., Patna, p 346

Sarawgi AK, Bisne R (2007) Studies on genetic divergence of aromatic rice germplasm for agro-morphological and quality characters. Oryza 44(1):74–76

Sarma KK, Ahmed T, Baruah DK (1990) Grain characteristics of some aromatic rice varieties of Assam. Intl Rice Res Newsl 15(1):13

Sharma N (2002) Quality characteristics of non-aromatic and aromatic (*Oryza sativa*) varieties of Punjab. Ind J Agric Sci 72(7):408–410

Shobharani N, Singh RK (2003) Efforts on aromatic rice improvement in India. In: Singh RK, Singh US (eds) A treatise on the scented rices of India. Kalyani Publishers, New Delhi, pp 23–72

Shobharani N, Subbarao LV, Viraktamath BC, Mishra B (2004) National guidelines for the conduct of tests for distinctness, uniformity and stability – Rice (*Oryza sativa* L.), DRR technical bulletin no 8. Directorate of Rice Research, Hydrabad

Shobharani N, Pandey MK, Prasad GSV, Sudharshan I (2006) Historical significance, grain quality features and precision breeding for improvement of export quality basmati varieties in India. Ind J Crop Sci 1(1–2):29–41

Siddiq EA, Vemireddy LR, Nagaraju J (2012) Basmati rices: genetics, breeding and trade. Agric Res 1(1):25–36

Singh VP (2000) The Basmati rice of India. In: Singh RK, Singh US, Khush GS (eds) Aromatic rices. Oxford and IBH Publishing, New Delhi, pp 135–154

Singh RK, Singh US, Khush GS, Rohilla R, Singh JP, Singh G, Shekhar KS (2000a) Small and medium grained scented rices of India. In: Singh RK, Singh US, Khush GS (eds) Aromatic rices. Oxford & IBH Pub. Co., New Delhi, pp 155–177

Singh RK, Gautam PL, Saxena S, Singh S (2000b) Scented rice germplasm: conservation, evaluation and utilization. In: Singh RK, Singh US, Khush GS (eds) Aromatic rices. Oxford & IBH Pub. Co., New Delhi, pp 107–133

Singh US, Rohilla R, Srivastava PC, Singh N, Singh RK (2003) Aromatic rices of Orissa. In: Singh RK, Singh US (eds) A treatise on the scented rices of India. Kalyani Publishers, New Delhi, pp 307–315

Singh N, Kaur L, Sodhi NS, Sekhon KS (2005) Physicochemical, cooking and textural properties of milled rice from different Indian rice cultivars. Food Chem 89:253–259

Aroma Volatile Characterization in Indian Scented Rice (*Oryza sativa* L.) Cultivars

4

4.1 Introduction

Flavor volatiles or aroma and texture are the principle sensory qualities of rice and have been rated as the major criteria for preference. Consumer's appreciation and the wide popularity of scented rice augmented its demand in domestic as well as in international markets. India holds the highest biodiversity of scented rice and has a huge number of basmati and nonbasmati rice varieties. Besides basmati type, some nonbasmati types – viz., Kalanamak, Sakerchini, and Hansraj grown in eastern Uttar Pradesh; Dubraj, Chinoor grown in Chhattisgarh; Kalajoha grown in northeastern states; Randhunipagal grown in Orissa; and Ambemohar grown in Maharashtra – are popularly cultivated and exported.

More than 100 volatiles contribute to rice aroma. Among these, 2-acetyl-1-pyrroline (2AP) possesses low odor threshold value; hence, it is regarded as principle aroma compound contributing to the aroma character of rice (Buttery et al. 1982, 1983). Various attempts were made to detect aroma in scented rice. In 1978, Sood and Siddiq (1978) developed a simple qualitative test to detect aroma in scented rice varieties. Nadaf et al. (2006) have developed a histochemical test to detect 2AP in scented rice. A marker system for validation of basmati types is developed at the Center for DNA Fingerprinting and Diagnostics (CDFD), Hydrabad, by Nagaraju et al. (2002). Even though India is one of the major exporters of basmati and nonbasmati rice, there is no method developed to quantify aroma volatiles.

4.1.1 Solid-Phase Microextraction (SPME) Method for Volatile Analysis

Recently, solid-phase microextraction (SPME) has emerged as a powerful analytical technique that can overcome the difficulties encountered in traditional methods. SPME was first described by Arthur and Pawliszyn (1990). SPME is a solvent-free method of extracting analytes from matrices by portioning them from a liquid or gaseous sample into an immobilized stationary phase (Steffen and Pawliszyn 1996).

A. Nadaf et al., *Scented rice (Oryza sativa L.) Cultivars of India: A Perspective on Quality and Diversity*, DOI 10.1007/978-81-322-2665-9_4

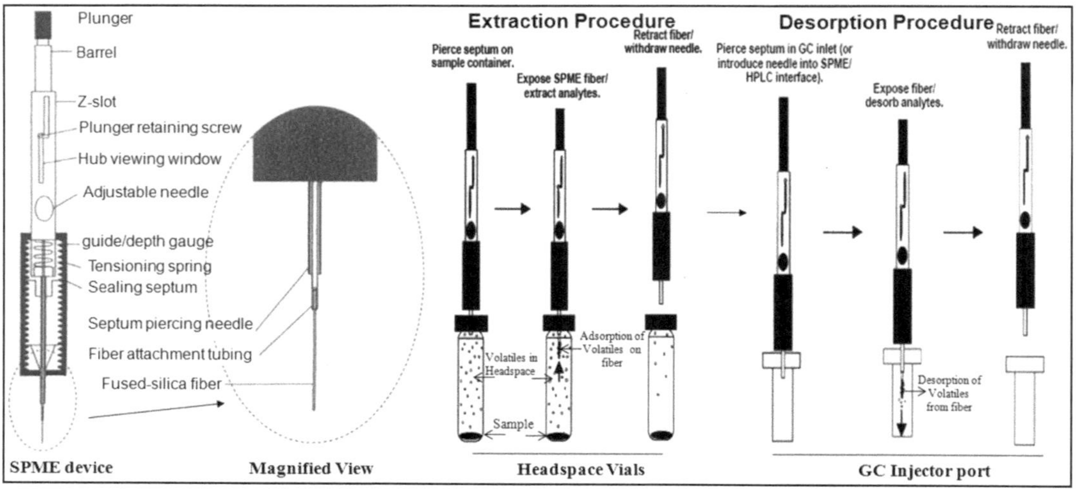

Schematic diagram of a commercial solid-phase microextraction device (Source: Supelco Bulletin 923, www.sigma-aldrich.com)

It is based on the partition of the analyte between extraction phase and matrix. The method uses a small fused silica fiber, coated with a suitable polymeric extracting phase, mounted in a syringe-like protective holder. When the SPME fiber is suspended in the headspace above the sample, the option is called headspace-SPME (HS-SPME). In HS-SPME, the volatiles released from a sample matrix in the headspace of vial are first adsorbed on the fiber in a sufficient amount of time and immediately desorbed in the injector port of analytical instrument (gas chromatography (GC)) for further characterization (Pico et al. 2007) (Fig. 4.1). HS-SPME enables faster sampling, extraction, concentration, sample introduction within an hour with high sensitivity and accuracy. The method saves time for preparation of sample and overall cost and improves detection limits and hence can be effectively used for analysis of aroma volatiles (Kataoka 2002). Moreover, the method is simple and sensitive and requires less sample (Stashenko and Martinez 2007; Hamm et al. 2005). In recent years, SPME has been successfully used to analyze 2AP from rice and *Vallaris glabra* flowers (Grimm et al. 2001; Wongpornchai et al. 2003; Laguerre et al. 2007 and Mathure et al. 2011).

In the present chapter, using solid-phase microextraction (SPME), a highly sensitive technique, a standard method is developed to analyze the aroma compounds qualitatively and quantitatively and 91 upland rice cultivars and 35 marketed rice varieties are assessed for their volatile contents.

4.2 Methodology and Approaches Adopted for Aroma Volatile Analysis

4.2.1 Optimization of Extraction Conditions

As SPME is based on equilibration between sample (matrix), headspace of vial and fiber, variation in extraction regimes affects the recovery of analyte. Hence, before quantification the HS-SPME conditions were varied and the corresponding variation in 2AP peak area was recorded to determine the optimum conditions.

4.2.1.1 Method for Extraction of 2AP and Other Volatiles Using HS-SPME

Cultivar Kamavatya – a scented landrace – was used for initial standardization of the method.

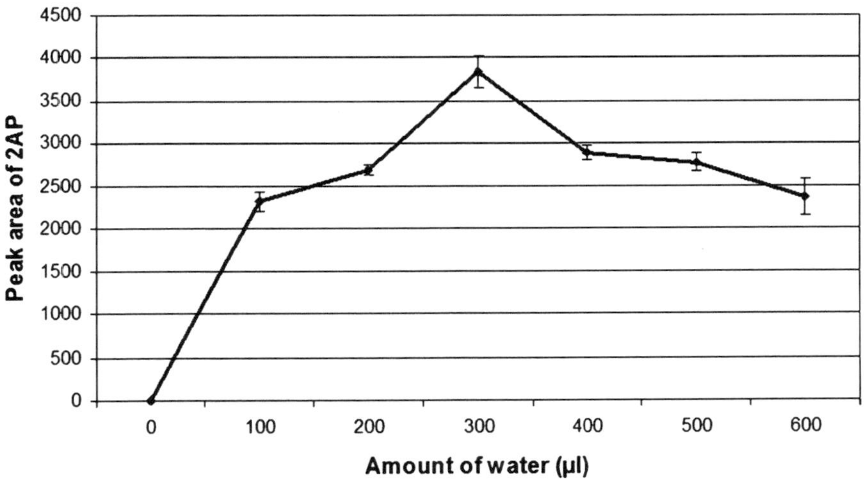

Fig. 4.1 Recovery of 2AP under varied amount of water added to rice sample (Source: Mathure et al. 2010)

The extraction was performed in 4-ml screw top vials (15×45 mm) with PTFE silicon septa (Chromatography Research Supplies, Louiseville, KY, USA). The vials were heated in oven set at 150 °C for 1 h prior to use to eliminate unintended volatile compounds. Fiber measuring 1 cm in length coated with Carboxen/Divinylbenzene/Poly-dimethyl-siloxane with manual holder was used for the extraction of 2AP (Grimm et al. 2001; Laguerre et al. 2007; Wercinski 1999). Authentic 2AP was a generous gift from Dr. P. Srinivas (Central Food and Technology Research Institute, Mysore, India). 2AP and randomly selected volatiles were identified using GC-MS (Shimadzu QP 5050A, Japan) with BP-20 capillary column (30 m × 0.32 µm). Separation and analysis of headspace volatiles from rice were performed using GC (Shimadzu 17 A, Japan), coupled with a flame ionization detector (FID). Volatiles were extracted and concentrated using SPME manual holder assembly (57330-U) equipped with SPME fiber (Supelco, Bellefonte, PA, USA) conditioned at 250 °C for 30 min. The fiber was desorbed at 250 °C injector temperature in splitless mode. The GC oven was programmed as 1-min hold at 50 °C, ramped to 100 °C at the rate of 4 °C/min and was further ramped to 240 °C at the rate of 50 °C/min with final hold of 2 min (Mathure et al. 2011). Optimization was carried out with respect to the sample weight, quantity of water, temperature of extraction, equilibration time, and adsorption time in triplicates. The weights of the sample analyzed were 0.5, 1, and 1.5 g. Quantity of water ranged from 0 to 600 µl with an increment of 100 µl; temperature of extraction ranged from 70 °C to 100 °C with an increment of 10 °C; pre-incubation time from 10 to 50 min with an increment of 10 min and adsorption time from 10 to 30 min with an increment of 5 min were used for analyzing the 2AP recovery. Area count was used as a measure of the quantity during optimization. The conditions were varied so as to achieve maximum increase in 2AP peak area (Mathure et al. 2011, 2014).

4.2.2 Assessment of Aroma Volatiles in Marketed Rice

4.2.2.1 Rice Samples

Thirty-three scented and two nonscented (Kolam Brand 2 and Manila) rice samples sold under various brands were procured from the supermarket (Pune, Maharashtra, India) for quantification of volatiles and assessing suitability of method. These samples were broadly classified as of basmati type (Basmati 370 and Basmati brand 1–10), ambemohar type (Ambemohar 157 and Ambemohar brand 1–4), kolam type (Kolam brand 1–4), indrayani type (Indrayani Brand 1 and 2), and other local types (Basumati, Chinoor, Dubraj, Ghansal, Kali kumud, Kalimuch, Kamavatya, Kamod, Khadkya, Kothmirsal, Lal dodki, Manila, and Raibhog).

4.2.2.2 Development of Calibration Curve for Quantification of 2AP

The calibration graph for quantification of 2AP was developed using optimized sampling conditions by standard addition approach. For this, 1-g sample of scented rice was processed following optimized conditions and peak area of 2AP was recorded as zero response. Rice was spiked with 1 μl dilution of standard 2AP. The vials were kept at 27 °C for 10-min equilibration prior to analysis. Due to limitation on quantity of standard 2AP, for higher concentrations rice was spiked with 10-mg pandan (*Pandanus amaryllifolius*) leaves containing known quantity of 2AP. Quantity of 2AP from pandan leaves was determined in triplicates using a standard method available in the laboratory (Wakte et al. 2010). Increase in the area of 2AP from 1-g rice and 1-g rice spiked with 2AP in triplicate was noted. These data were used to derive calibration curve of 2AP.

4.2.2.3 Development of Calibration Curve for Quantification of Aroma Volatiles Other than 2AP

Analytical grade standards of hexanal, nonanal, benzyl alcohol, indole, vanillin (Aldrich, Steinheim, Germany), guaiacol (Fluka, Steinheim, Germany), and decanal (Sigma, Steinheim, Germany) were used for peak identification and quantification. The calibration graph for quantification of these volatiles was developed using optimized sampling conditions by standard addition approach. For this, 1-g sample of scented rice was processed following optimized conditions and peak areas for compounds under study were recorded as zero response. The same rice stock was used to develop calibration curves. Rice was spiked with 1 μl spike mix containing known concentration of hexanal, nonanal, decanal, benzyl alcohol, vanillin, guaiacol, and indole. The analysis was performed in triplicates. Increase in the area of volatiles over 1-g rice to that of rice spiked with volatiles was used to derive standard curve for respective volatile.

4.2.2.4 Assessment for Suitability of Method Using Marketed Rice Samples

The optimized conditions were validated using 35 marketed rice samples for quantification of 2AP and other aroma volatiles. To ensure optimum performance of SPME fiber employed in the analysis, it was checked after every ten samples with the standard fiber maintained separately. The only fiber having performance comparable with the standard fiber was continued in further analysis. The new fiber was also checked for the optimum performance before employing it in the analysis of rice samples. Blank run of GC was also performed after every ten samples to remove traces (if any) from earlier runs in GC column (Mathure et al. 2011).

4.2.2.5 Data Analyses

Mean values for each volatile compound within the rice categories were compared using Duncan's multiple range test (DMRT). Principle component analysis (PCA) was done to visualize the differences among the volatile compositions in rice samples. Pearson's correlation coefficients (r) for PCs and rice types were also estimated using SPSS software (version 11.5, Chicago, USA).

4.2.2.6 Assessment of Aroma Volatiles in Upland Rice Cultivars

The 91 scented rice cultivars cultivated in upland regions were analysed for their aroma contents.

4.2.2.7 Rice Samples

Ninety-one rice (*Oryza sativa* L. spp *indica*) cultivars representing nonbasmati scented from Maharashtra state, nonbasmati scented from Karnataka state, nonbasmati scented from other states of India, basmati and nonscented rice cultivars were selected for the study (Table 4.1). The cultivars were classified as landrace (indigenous cultivar), selection (pure line selection from land-

Table 4.1 Details of upland rice (*Oryza sativa* L.) cultivars used in analysis the of headspace volatiles

Category	State	Type	Cultivar
Nonbasmati scented (77)	*Maharashtra (31)*	Landrace	Ambemohar, Ambemohar Ajra, Ambemohar Pandhara, Ambemohar-Tambda, Champakali, Chimansal, Gham, Ghansal, Girga, Jiri, Kala bhat, Kalsal, Kamavatya, Kamod, Kate chinoor, Khadkya, Kondhekar chinoor, Kothmirsal, Lal bhat, Lal dodki, Parabhani chinoor, Raibhog, Tamsal, Velchi
		Selection	Ambemohar-157, RDN scented, RDN local
		Hybrid	Bhogavati, Indrayani, Pawana, Phule radha
	Karnataka (13)	Landrace	Basumati, Gandhesale, Geerige sanna, Kagisali, Kali kumud, Kaligajvili, Kumud, Medhini sanna bhata, Mysore sanna, Sanna bili bhata, Vasane sanna bhatta
		Hybrid	Makarand, Mugad sugandha
	Other states (33)	Landrace	Acharmati, Amritbhog, Badsahbhog, Bansaphool A, Barke bhat, Bela blue, Bishnubhog, Dhanaprasad, Dubraj, Dubrajsena, Dusara, Elaichi, Gatia, Girija samba, Jeeraphool, Jeera-sona, Kalakrishna, Kalajeera, Kalanamak, Kanakjeer, Kothmbiri, Lalu, Pakhe bhat, Pimpudibasa, Prabhatjeera, Ratibhog, Shrabanmasi, Shyamjeer, Tulshimanjula, Tulsiamrit, Tulsikanti, Velkat
		Hybrid	Pusa sugandha
Basmati (9)	*Other states*	Landrace	Kernal local, Pakistan basmati
		Selection	Basmati 370, Basmati 376, Basmati 386, Basmati 6311
		Hybrid	Pusa basmati, Pusa basmati-1, Super basmati
Nonscented (5)	*Maharashtra*	Landrace	Kolamb
		Hybrid	Jaya, Sonsali
	Karnataka	Landrace	Chitak bhat, Manila

Source: Mathure et al. 2014

race), and hybrid (developed by crossbreeding). These rice cultivars were available at the Dept. of Botany, University of Pune, Pune, Maharashtra, India, either as a personal collection from Maharashtra and Karnataka or as a germplasm procured from rice research stations and institutes, viz., Agricultural Rice Research Station, Radhanagari (Kolhapur, Maharashtra, India); Rice Research Station, Shindevahi (Chandrapur, Maharashtra, India); Rice Research Station, Karjat (Raigadh, Maharashtra, India); Agricultural Research Station, Sirsi (Karwar, Sirsi, Karnataka, India); Rice Research Station, Mugad (Dharwad, Karnataka, India); Indian Agricultural Research Institute (New Delhi, India); and National Seed Corporation Ltd (New Delhi, India). The seedlings of these cultivars were raised in the experimental field at Karjat Rice Research Station, Karjat, District Raigadh, Maharashtra state, India, in Kharif 2009, and seeds were harvested following routine cultivation practice (Mathure et al. 2014). The paddy was dehusked prior to analysis.

4.2.2.8 Extraction, Identification, and Quantification of Volatiles

The volatiles were extracted following optimized SPME conditions. Besides 2AP, hexanal, nonanal, decanal, benzyl alcohol, vanillin, guaiacol, and indole, 15 other volatiles, viz., pentanal (Merk, Hohenbrunn, Germamy); octanal, *trans*-2-octenal, 1-tetradecene, *trans*-2-nonenal, 2-phenylethanol, nonanoic acid, 2 amino acetophenone (Sigma-Aldrich, St. Louis, MO, USA); heptanal, *trans*-3-octen-2-one, 1-octen-3-ol, (*E*,*E*)-nona-2,4-dienal (SAFC Supply Solutions, St. Louis, MO, USA); 1-hexanol (Supelco Analytical, PA, USA); 4 vinyl guaiacol, 4 vinyl phenol (Alfa Aesar, Karlsruhe, Germany) were used for peak identification and quantification.

The calibration graph for quantification of these volatiles was developed using optimized sampling conditions by standard addition approach. Rice was spiked with 1 μl spike mix containing known concentration of 15 volatiles. The analysis was performed in triplicates. Increase in the area of volatiles over 1-g rice to that of rice spiked with volatiles was used to derive standard curve for respective volatile. Dilutions of standards were made in methanol (Merck, Mumbai, India).

All the 91 cultivars were subjected to optimized extraction conditions to assess 23 volatiles quantitatively.

4.2.2.9 Estimation of Odor Active Values

Odor active values (OAVs) were calculated by dividing quantity of compound by the reported odor threshold value of that compound.

4.2.2.10 Data Analyses

Descriptive analysis of volatiles to determine average quantity, range of volatile amount, and % coefficient of variation was performed. Duncan's multiple rage test (DMRT) was performed on average values for each volatile compound within rice categories (nonbasmati scented from Maharashtra state, nonbasmati scented from Karnataka state, nonbasmati scented from other states of India, basmati, and nonscented) to identify compounds exhibiting significant variation within categories. Pearson's correlation coefficients (r) among compounds were estimated. Principle component analysis (PCA) was performed to study variation in composition of volatiles among cultivars. All the analysis was performed using SPSS software (version 11.5, Chicago, USA).

4.3 Profiles of Aroma Volatile Analysis in Indian Scented Rice Cultivars

4.3.1 Optimization of HS-SPME Conditions for Quantification of 2AP in Rice

Purity of authentic 2AP was confirmed by GC–MS QP-5050A (Shimadzu, Kyoto, Japan) with Rtx-5 capillary column (60 m × 0.25 μm, Restek Corporation, Bellefonte, PA, USA). The mass spectrum showed major ions 41, 42, 43(100), 68,

69, 83, 111, which were identical to the ions of 2AP in NIST147 library and also with the ions mentioned in existing literature (Buttery et al. 1983; Mahatheeranont et al. 2001). This standard was used for identification of peak and quantification.

4.3.1.1 Optimization for Sample Weight

The amount of 2AP measured by peak area showed a higher variation with 0.5 g when compared to 1 g of sample. Sample amount of 1.5 g was not suitable for SPME because of inadequate vial headspace. Therefore, 1-g sample was used for further analyses.

4.3.1.2 Optimization for Water Quantity

Addition of water to rice kernel allows optimum extraction of 2AP (Grimm et al. 2001; Laguerre et al. 2007). In the present study, in absence of water, 2AP was not detected. This could be due to the use of scented landrace Kamavatya instead of Jasmine (Grimm et al. 2001) and Basmati (Laguerre et al. 2007) rice. 2AP content of Kamavatya may be very low compared to that of Jasmine and Basmati rice. However, marked increase in 2AP area was recorded with the addition of water up to 300 μl (Fig. 4.1). Partial starch gelatinization with water addition reduces kernel hardness, and this might have helped to release aroma compounds. Further addition of water (>300 μl) to kernels resulted in reduction of 2AP. Addition of water would have affected the equilibrium of 2AP in matrix, headspace, and fiber, thus resulting in decreased 2AP area. Hence, the results indicated that the amount of aroma released by the addition of water and its partition within matrix and headspace plays a critical role in the optimization of water addition in the sample (Mathure et al. 2011). Grimm et al. (2001) and Laguerre et al. (2007) have recorded such increased 2AP recovery over a narrow range of water added during extraction.

4.3.1.3 Optimization of Extraction Temperature

Variations in equilibrium of volatile or sensitivity of volatile compounds to increased temperature are known to attribute to temperature-dependent variation. Extraction temperature was varied from 70 to 100 °C to determine optimum temperature giving maximum 2AP recovery, and the results are shown in Fig. 4.2. Relative area of 2AP increased 1.4 times with increase in temperature from 70 to 80 °C (Fig. 4.2). Marginal nonsignificant increase in 2AP peak area at 80 °C over 90 °C was recorded (Mathure et al. 2011).

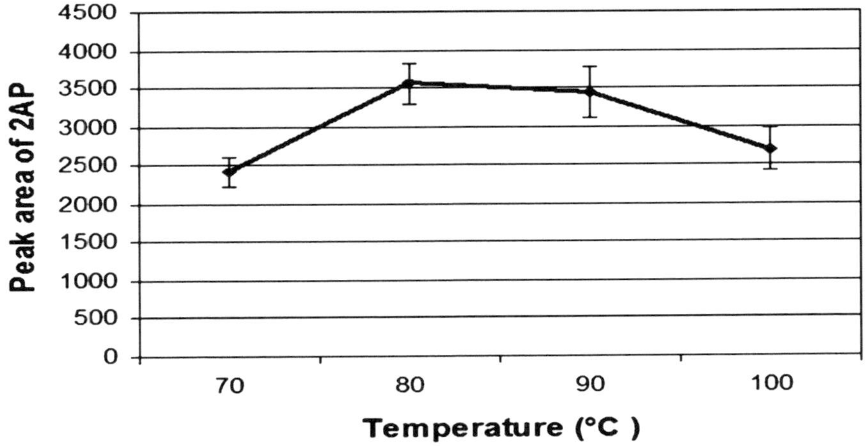

Fig. 4.2 Recovery of 2AP under varied extraction temperatures

Further increase in temperature reduced 2AP peak area. Hence, 80 °C was taken as the optimum temperature of extraction. Similar observations were recorded by Grimm et al. (2001).

4.3.1.4 Optimization of Extraction Time

Generally, a preincubation time of 10–15 min is reported to be sufficient for optimal extraction of most volatile components (Grimm et al. 2001). Analysis was carried out by varying the preincubation and keeping a 15-min extraction time (Fig. 4.3). The 2AP peak area was found to vary with the preincubation time and was maximum at 30 min (Mathure et al. 2011).

Further, the analyses were carried out by preincubating for 30 min and varying the adsorption time, and the results are shown in Fig. 4.4. Maximum 2AP area was recorded at 20-min adsorption time. Thus, time variations in extraction period considerably affected 2AP recovery (Mathure et al. 2011). Zeng et al. (2009) recorded variation in composition of volatiles during cooking. The longer extraction period could have reduced the adsorption of 2AP over other volatiles resulting in decreased 2AP peak area.

Extraction at 80 °C for 30-min preincubation followed by 20-min adsorption from 1-g rice containing 300 μl of odor-free water were the optimum conditions.

4.3.2 Assessment of Aroma Volatiles in Marketed Rice

4.3.2.1 Development of Calibration Curve for 2AP and Seven Other Aroma Volatiles

On the basis of literature, seven volatiles besides 2AP were selected. The analytical grade GC standards of these volatiles were used for identification of peak and quantification of volatile. Chromatograph of rice headspace volatiles obtained by GC-FID is presented as Fig. 4.5. Each volatile has different extent of affinity for the SPME fiber during extraction and varies in amount present in rice. So depending on these factors, concentration of spike mix and number of dilutions for construction of standard graph were decided. Increase in area of each peak with respect to zero response (without spike mix) was recorded. Calibration curves for volatiles were plotted with a linear correlation coefficient approaching 1 (Figs. 4.6 and 4.7).

4.3.2.2 Quantification of Aroma Volatiles in Marketed Rice

Quantitative analysis of 2AP and other volatiles in 35 rice samples was carried out by method developed (Table 4.3).

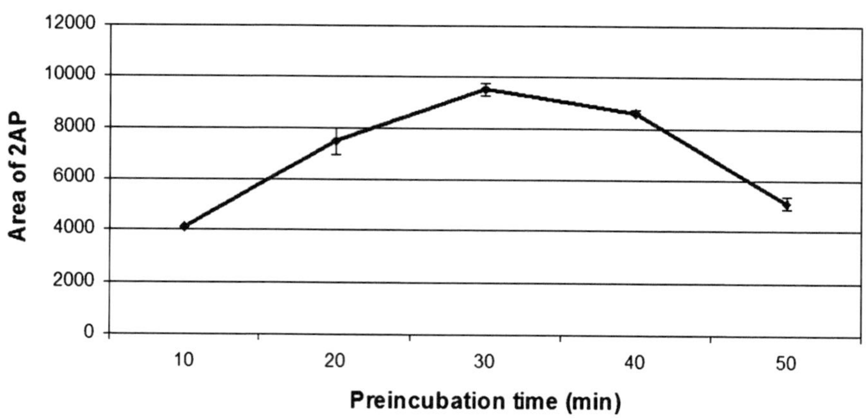

Fig. 4.3 Recovery of 2AP under varied preincubation time (Source: Mathure et al. 2010)

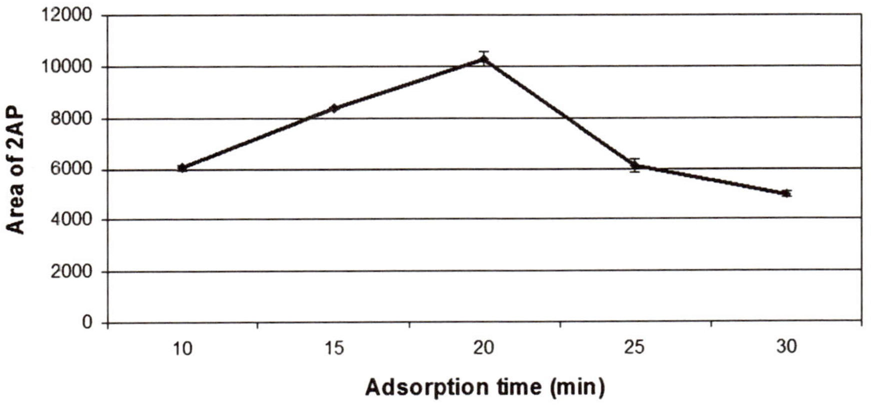

Fig. 4.4 Recovery of 2AP under varied adsorption time (Source: Mathure et al. 2010)

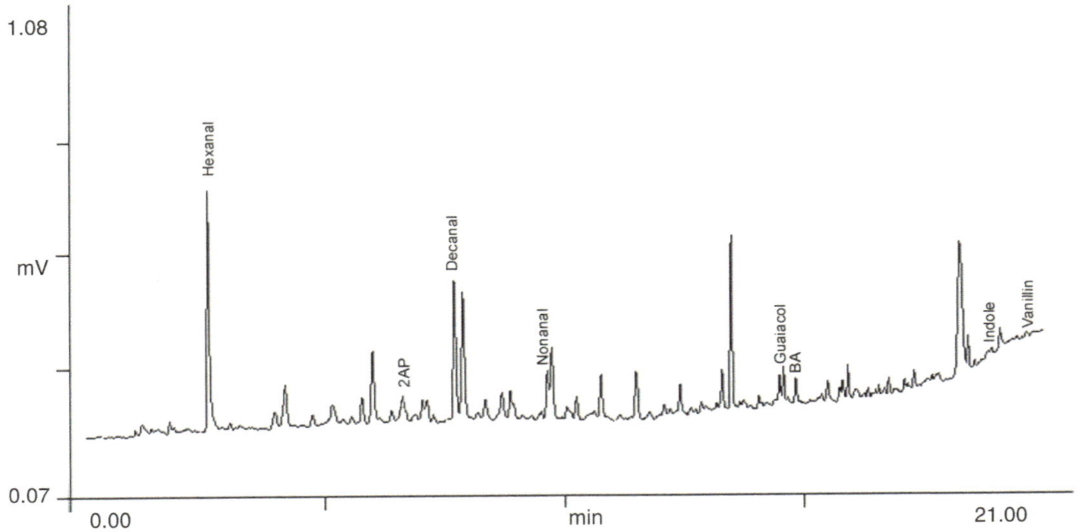

Fig. 4.5 GC-FID chromatograph of rice headspace volatiles

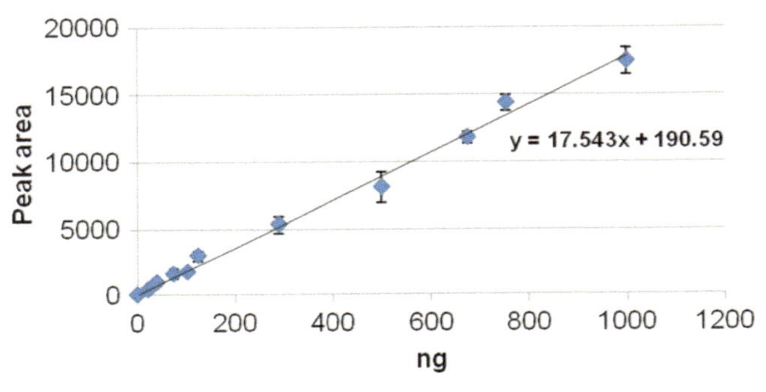

Fig. 4.6 Standard graph for quantification of 2AP

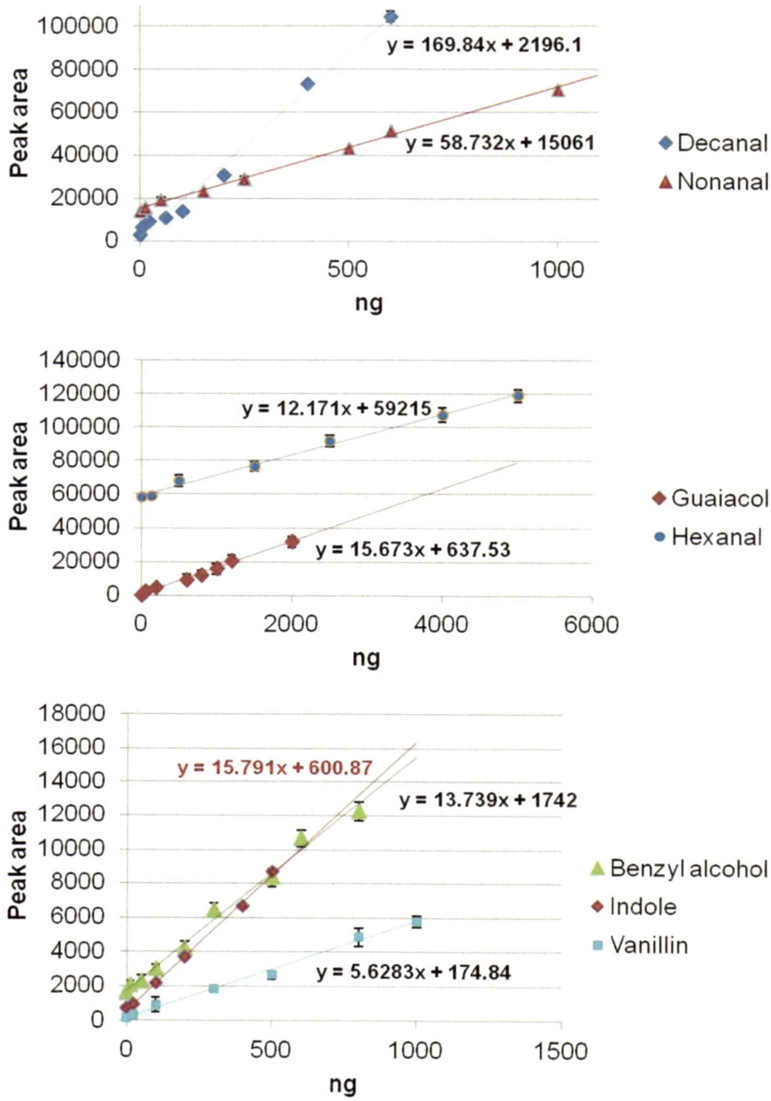

Fig. 4.7 Standard graphs for quantification of seven volatiles

Quantification of 2AP

Indrayani Brand 2 recorded the highest amount of 2AP (0.552 mg/kg), followed by Kamod (0.418 mg/kg) and Basmati Brand 5 (0.411 mg/kg). Least amount of 2AP was found in Kolam Brand 2 (0.032 mg/kg). Basmati types had 2AP content ranging from 0.122 mg/kg to 0.411 mg/kg. Among ambemohar types, significantly high 2AP content was found in Ambemohar Brand 3 and Brand 4 (0.344 mg/kg and 0.365 mg/kg, respectively). Ambemohar-157 recorded 2AP content (0.115 mg/kg) significantly lower than four other brands of ambemohar. Indrayani types exhibited variation in 2AP content (Table 4.3). Kolam types revealed variation in 2AP content from as low as 0.032 mg/kg in Kolam Brand 2 to 0.151 mg/kg in Kolam Brand 3. Kolam Brand 2 (0.032 mg/kg) and Kolam Brand 1 (0.079 mg/kg) perceived as nonscented and mild scented samples, respectively, revealed less 2AP content. In 13 other rice types, Manila was the only nonscented cultivar (0.064 mg/kg). It was found that Kamod – a landrace cultivated in the interiors of Nashik district of Maharashtra state – is of a strongly scented type (0.418 mg/kg). The present study revealed that Basumati, Ghansal, Kali

Table 4.2 Odor description and odor threshold values of ten major volatiles in scented rice (*O. sativa* L.) (Source: Mathure et al. 2010)

Retention time (min)	Volatile	Odor description	Odor threshold (μg/kg)
1.88	Pentanal	Nutty, sweet	12
2.72	Hexanal	Green, grasslike, herbal	4.5
4.13	Heptanal	Floral	3
5.99	Octanal	Slightly fruity, citrus like	0.7
6.98	2-acetyl-1-pyrroline	Cooked rice, sweet, pleasant, popcorn like	0.1
7.5	1-hexanol	Vegetal, green	2500
8.10	Nonanal	Flour, sticky rice, fatty, floral, fruity	1
18.15	Nonanoic acid	Animal, cheese	3000
19.34	4 vinyl phenol	Phenolic, medicinal	10
20.58	Vanillin	Vanilla odor	58

kumud, Kamod, Khadkya, Raibhog grown in Maharashtra state and parts of Karnataka state excel in 2AP content over some of the Basmati samples, Dubraj, Kalimuch, and Chinoor (Mathure et al. 2011).

Content of 2AP in milled Basmati was reported to vary in the following amounts: 0.06 mg/kg (Buttery et al. 1983), 0.588 mg/kg (Tava and Bocchi 1999), and 0.061 mg/kg (Nadaf et al. 2006) using steam distillation; 0.019–0.342 mg/kg (Bergman et al. 2000) and 0.235 mg/kg (Yoshihashi 2002) by solvent extraction; and 0.26–0.38 mg/kg by static headspace-GC (Sriseadka et al. 2006). Maraval et al. (2008) recorded 2AP content up to 0.264 mg/kg among scented cultivars, viz., Aychade, Fidji, and Giano. Our values are in agreement with these reports. However, lower values of 2AP content were reported in black rice (Yang et al. 2008a) and in six rice flavor types (Yang et al. 2008b) using Tenax trap. These studies expressed 2AP content in terms of either δ-carvone or 2, 4, 6-trimethylpyridine equivalent and lack exact content of 2AP (Mathure et al. 2011).

Quantification of Other Volatiles

Analysis of volatiles other than 2AP revealed that hexanal (0.787–7.528 mg/kg), nonanal (0.157–0.685 mg/kg), decanal (0.017–0.345 mg/kg), and benzyl alcohol (0.020–0.535 mg/kg) were present in all rice samples. Guaiacol was not detected in Indrayani Brand 1, Kolam Brand 4, Basumati, Ghansal, Lal dodki, and Manila. Indrayani Brand 2 recorded the highest guaiacol content (0.464 mg/kg). Indole was detected in one sample (Basmati Brand 9) of basmati and six samples of local rice types. Maximum indole (0.083 mg/kg) was recorded in Dubraj. Vanillin was present in 24 samples, majority of which were Basmati, Indrayani, and other local rice types. Vanillin was recorded in Ambemohar-157 and Kolam Brand 2.

Among basmati types, Basmati Brand 7 recorded 6.828 mg/kg hexanal. Basmati 370 recorded lowest content of hexanal (0.787 mg/kg), nonanal (0.242 mg/kg), decanal (0.065 mg/kg), guaiacol (0.033 mg/kg), and benzyl alcohol (0.052 mg/kg). However, it contained the highest amount of vanillin (0.324 mg/kg) among basmati types. With the exception of Basmati Brand 9 (0.021 mg/kg), indole was absent in basmati types.

Ambemohar Brand 2 recorded the highest hexanal content (7.528 mg/kg). Ambemohar-157 characteristically recorded significantly lower content of nonanal (0.173 mg/kg), decanal (0.022 mg/kg), guaiacol (0.049 mg/kg), and benzyl alcohol (0.049 mg/kg) among the ambemohar types. Moreover, only in this Ambemohar sample, vanillin (0.126 mg/kg) was detected (Mathure et al. 2011).

Development of Cultivar Specific Volatile Profile

Based on odor threshold value, ten aroma volatiles, including 2AP, were selected to develop a cultivar specific profile (Table 4.2). These ten volatiles were compared among the cultivars under study, and volatile profiles were developed.

Table 4.3 Quantification of 2AP and other volatile compounds from marketed rice samples (Source: Mathure et al. 2010)

Sr. No.	Types/Brands	Quantity (mg/kg)							
		2AP	Hexanal	Nonanal	Decanal	Guaiacol	Benzyl alcohol	Indole	Vanillin
Basmati types									
1	Basmati 370	0.214^{d}	0.787^{a}	0.242^{a}	0.065^{a}	0.033^{a}	0.052^{a}	nd^{a}	0.324^{g}
2	Basmati Brand 1	0.178^{c}	2.257^{b}	0.352^{c}	0.306^{e}	0.203^{d}	0.233^{fg}	nd^{a}	0.155^{f}
3	Basmati Brand 2	0.347^{f}	2.845^{cd}	0.353^{c}	0.264^{d}	0.276^{e}	0.176^{de}	nd^{a}	0.088^{d}
4	Basmati Brand 3	0.294^{e}	2.865^{cd}	0.242^{a}	0.220^{bc}	0.104^{b}	0.104^{b}	nd^{a}	0.068^{cd}
5	Basmati Brand 4	0.122^{a}	3.016^{d}	0.327^{bc}	0.241^{bcd}	0.164^{c}	0.243^{g}	nd^{a}	0.113^{e}
6	Basmati Brand 5	0.411^{g}	2.624^{c}	0.316^{b}	0.212^{b}	0.118^{b}	0.159^{cd}	nd^{a}	0.026^{b}
7	Basmati Brand 6	0.178^{c}	6.540^{h}	0.470^{e}	0.345^{f}	0.347^{g}	0.211^{f}	nd^{a}	0.083^{cd}
8	Basmati Brand 7	0.151^{b}	6.828^{h}	0.410^{d}	0.245^{cd}	0.170^{c}	0.186^{e}	nd^{a}	0.065^{cd}
9	Basmati Brand 8	0.196^{cd}	3.686^{e}	0.326^{bc}	0.314^{e}	0.305^{f}	0.170^{cde}	nd^{a}	nd^{a}
10	Basmati Brand 9	0.202^{d}	4.122^{f}	0.426^{d}	0.323^{ef}	0.374^{h}	0.178^{de}	0.021^{b}	0.087^{d}
11	Basmati Brand 10	0.194^{cd}	6.230^{g}	0.416^{d}	0.260^{d}	0.203^{d}	0.152^{c}	nd^{a}	0.062^{c}
Ambemohar types									
12	Ambemohar-157	0.115^{a}	4.509^{a}	0.173^{a}	0.022^{a}	0.049^{a}	0.049^{a}	nd	0.126^{b}
13	Ambemohar Brand 1	0.313^{c}	5.605^{c}	0.380^{c}	0.240^{d}	0.279^{d}	0.207^{c}	nd	nd^{a}
14	Ambemohar Brand 2	0.204^{b}	7.528^{d}	0.260^{b}	0.215^{c}	0.175^{c}	0.201^{bc}	nd	nd^{a}
15	Ambemohar Brand 3	0.344^{d}	5.011^{b}	0.349^{c}	0.193^{b}	0.330^{e}	0.255^{d}	nd	nd^{a}
16	Ambemohar Brand 4	0.365^{d}	4.353^{a}	0.268^{b}	0.217^{c}	0.115^{b}	0.189^{b}	nd	nd^{a}
Kolam types									
17	Kolam Brand 1	0.079^{b}	2.015^{a}	0.243^{a}	0.252^{d}	0.134^{b}	0.212^{c}	nd	nd^{a}
18	Kolam Brand 2	0.032^{a}	2.357^{b}	0.475^{c}	0.160^{b}	0.169^{c}	0.133^{a}	nd	0.053^{b}
19	Kolam Brand 3	0.151^{c}	2.552^{b}	0.296^{b}	0.226^{c}	0.177^{c}	0.161^{b}	nd	nd^{a}
20	Kolam Brand 4	0.139^{c}	2.381^{b}	0.269^{ab}	0.084^{a}	nd^{a}	0.535^{d}	nd	nd^{a}

Indrayani types									
21	Indrayani Brand 1	0.314	4.193	0.210	0.027	nd	0.020	nd	0.044
22	Indrayani Brand 2	0.552	3.426	0.380	0.260	0.464	0.204	nd	0.177
Other local types									
23	Basumati	0.190^{d}	1.587^{b}	0.239^{c}	0.065^{b}	nd^{a}	0.202^{f}	nd^{a}	0.150^{c}
24	Chinoor	0.137^{bc}	2.673^{e}	0.241^{c}	0.203^{c}	0.376^{f}	0.193^{f}	0.067^{d}	0.273^{d}
25	Dubraj	0.124^{bc}	2.493^{d}	0.251^{c}	0.219^{f}	0.388^{f}	0.094^{c}	0.083^{e}	0.333^{e}
26	Ghansal	0.237^{f}	1.581^{b}	0.335^{d}	0.060^{b}	nd^{a}	0.032^{a}	nd^{a}	nd^{a}
27	Kali kumud	0.216^{e}	1.767^{c}	0.198^{b}	0.093^{c}	0.095^{d}	0.266^{g}	nd^{a}	nd^{a}
28	Kalimuch	0.123^{b}	2.406^{d}	0.234^{c}	0.201^{e}	0.414^{g}	0.113^{d}	nd^{a}	0.249^{d}
29	Kamavatya	0.146^{c}	3.587^{f}	0.157^{a}	0.017^{a}	0.023^{bc}	0.055^{b}	nd^{a}	0.145^{c}
30	Kamod	0.418^{g}	4.368^{gh}	0.417^{e}	0.165^{d}	0.169^{e}	0.058^{b}	0.025^{b}	0.073^{b}
31	Khadkya	0.198^{de}	4.520^{h}	0.207^{b}	0.028^{a}	0.011^{ab}	0.047^{ab}	0.033^{b}	0.029^{a}
32	Kothmirsal	0.144^{bc}	1.811^{c}	0.251^{c}	0.088^{c}	0.016^{bc}	0.161^{e}	0.065^{d}	0.143^{c}
33	Lal dodki	0.177^{d}	1.092^{a}	0.685^{f}	0.067^{b}	nd^{a}	0.113^{d}	nd^{a}	0.070^{b}
34	Manila	0.064^{a}	1.674^{bc}	0.195^{b}	0.055^{b}	nd^{a}	0.201^{f}	nd^{a}	0.251^{d}
35	Raibhog	0.253^{f}	4.307^{g}	0.242^{c}	0.025^{a}	0.032^{c}	0.059^{b}	0.045^{c}	nd^{a}
	Min	**0.032**	**0.787**	**0.157**	**0.017**	**nd**	**0.020**	**nd**	**nd**
	Max	**0.552**	**7.528**	**0.685**	**0.345**	**0.464**	**0.535**	**0.083**	**0.333**
	Average	0.215	3.417	0.310	0.171	0.163	0.161	0.010	0.091
	CV%	51.93	49.48	34.66	58.74	87.28	59.26	227.37	107.33

Values with the same letter superscripted in each rice type for each volatile are not significantly different at $p=0.05$
nd not detected

Among four samples of Kolam, hexanal content varied in a narrow range (2–2.6 mg/kg). In Kolam Brand 4, decanal was significantly low (0.084 mg/kg) and benzyl alcohol was significantly high (0.535 mg/kg) than other samples in this class. In kolam types, with the exception of Kolam Brand 4, guaiacol was detected in other samples and vanillin was detected only in Kolam Brand 2 (0.053 mg/kg).

Indrayani types exhibited higher extent of variation in content between samples for all volatiles assessed. Indrayani Brand 2 recorded maximum content of nonanal (0.380 mg/kg), decanal (0.260 mg/kg), benzyl alcohol (0.204 mg/kg), and vanillin (0.177 mg/kg), whereas Indrayani Brand 1 recorded minimum content of nonanal (0.210 mg/kg), decanal (0.027 mg/kg), benzyl alcohol (0.020 mg/kg), and vanillin (0.044 mg/kg). Even though Indrayani Brand 2 revealed a maximum content of guaiacol (0.464 mg/kg), it was absent in Indrayani Brand 1. In other local rice types, indole was recorded in Kamod, Khadkya, Raibhog, Kothmirsal, Chinoor, and Dubraj (0.025–0.083 mg/kg). In this group, Khadkya, Lal dodki, Kamod, Kothmirsal, Kamavatya, Basumati, Kalimuch, Manila, Chinoor, and Dubraj recorded vanillin (0.070–0.333 mg/kg). Guaiacol attributing to smoky odor was detected in low amount (<0.032 mg/kg) in Khadkya, Kothmirsal, and Raibhog and in moderate amount (<0.169 mg/kg) in Kali kumud and Kamod. Higher content of guaiacol was recorded in Chinoor (0.376 mg/kg), Dubraj (0.388 mg/kg), and Kalimuch (0.414 mg/kg).

Among the volatiles, hexanal has been identified as a volatile giving off-odor to rice (Bergman et al. 2000). Bergman et al. (2000) reported 0.543–2.209 mg/kg, and Tava and Bocchi (1999) reported 3.238 mg/kg of hexanal in Basmati. As far as other volatiles are concerned, Tava and Bocchi (1999) reported nonanal (0.065 mg/kg), decanal (0.075 mg/kg), benzyl alcohol (0.187 mg/kg), and indole (0.287 mg/kg) in Basmati. Maraval et al. (2008) detected hexanal (42–122 μg/kg), nonanal (134–188 μg/kg), decanal (61–107 μg/kg), guaiacol (traces), indole (69–365 μg/kg), and vanillin (139–399 μg/kg) in three scented rice. Our values are in agreement with these reports; however, we obtained higher amount of nonanal in all samples and indole was detected in only one of the basmati types in less amount than the reported values. Studies by Yang et al. (2008a, b) recorded hexanal, nonanal, decanal, guaiacol, and indole content in rice. However, similar to that of 2AP, contents of other volatiles are also less than the values obtained in the present study.

In the present study, benzyl alcohol (odor threshold 10 mg/kg in water) and indole content (odor threshold 0.14 mg/kg in water) were recorded below their odor thresholds. The results are in agreement with Buttery et al. (1988), which indicated that benzyl alcohol and indole have no contribution to the aroma of cooked rice. Content of hexanal, nonanal, decanal, and guaiacol in rice was higher than their odor threshold, thus contributing to aroma. It was observed that vanillin content in 21 samples was above its odor threshold (0.058 mg/kg), thus contributing to the aroma of these samples. Vanillin has been reported as an important flavor ingredient in Basmati and small grained rice (Jezussek et al. 2002), cooked rice (Maraval et al. 2008), and glutinous rice (Zeng et al. 2009). Our study confirms their findings and point out the role of vanillin as an important volatile compound in rice aroma and flavor besides 2AP (Mathure et al. 2011).

4.3.3 Principle Component Analysis

In the PCA analysis for eight volatiles, first three components (PCs) explained 30.32, 20.93, and 14.71 % of the total variance (Fig 4.8). The rice types exhibited significant correlation ($r = -0.598$, at $p = 0.01$) with PC1, indicating that each rice type contributed up to 30.32 % variability in volatile composition. Within the rice types, PC2 and PC3 further separated the cultivars of basmati and local types, which recorded higher spread than that of ambemohar and kolam types (Fig. 4.9).

In India, Ambemohar has been marketed as a popular variety, fetching 15–20 % higher market price than Kolam, owing to its aroma. In the PCA, ambemohar was placed closer to basmati

Fig. 4.8 PCA plot of eight volatile compounds (Source: Mathure et al. 2010)

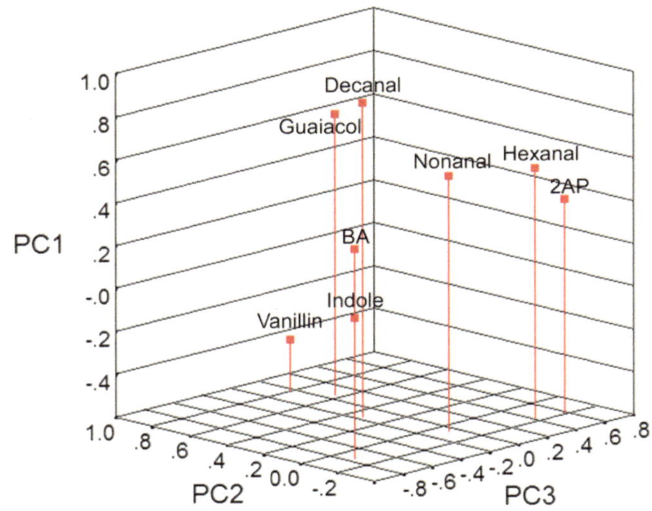

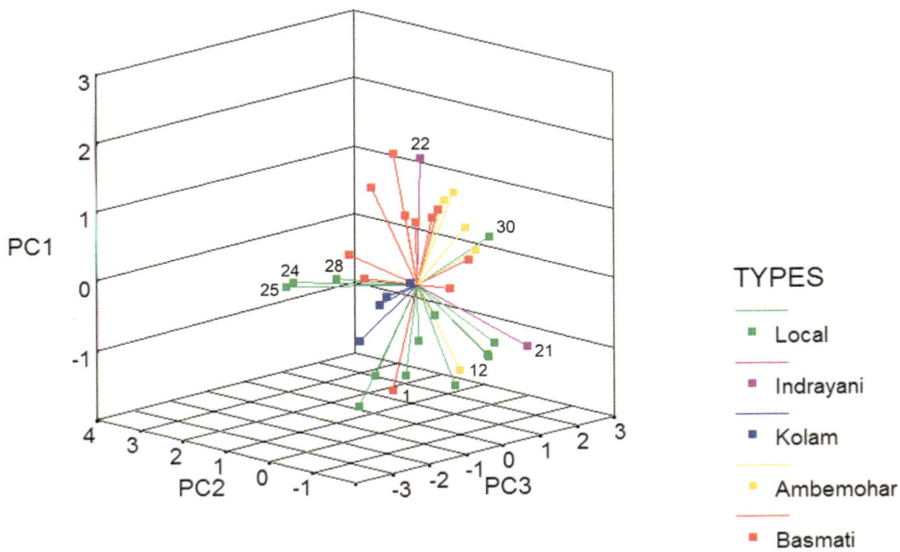

Fig. 4.9 Score plot showing distribution of 35 rice samples (number corresponds to the rice sample in Table 4.3) (Source: Mathure et al. 2010)

types, indicating comparable volatile profile with that of basmati.

The score plots separated Basmati 370 and Ambemohar-157 from basmati and ambemohar types respectively. The indrayani samples were also placed distantly. The variation in volatile composition within rice types could be due to variation in locality of cultivation, postharvest processing, or storage conditions in the market. Chinoor, Dubraj, Kalimuch, and Kamod segregated from other local rice types. Ambemohar types and Kamod were closer to basmati types than Chinoor, Dubraj, and Kalimuch (Mathure et al. 2011).

The present study revealed that the content of 2AP and other volatiles in Kamod, Raibhog, Ghansal cultivars and ambemohar types are comparable to Basmati, Dubraj, and Kalimuch

(Table 4.3, Fig. 4.9). Basmati, Dubraj, Kalimuch, and Chinoor seize higher market share owing to their aroma and taste. Thus, the study highlights the potential and marketability of local cultivars Kamod, Raibhog, Ghansal that could be explored further to boost their popularity among consumers (Mathure et al. 2011).

Our study confirms the utility of SPME calibrated with standard addition for rapid quantification of aroma volatiles from 1 g of sample within 50 min per sample. The protocol developed for 2AP can also be applied for quantification of other volatiles. Most recently, Bradbury et al. (2005) raised a serious concern about the accurate assessment of 2AP in breeding high-yielding fragrant rice cultivars and selection of the subtle recessive trait of fragrance within individual plants. This method is easy to automate and does not require organic solvents for extraction (Mathure et al. 2011).

Owing to the sensitivity of the method developed, it can be effectively employed in the quantification of aroma on a commercial scale, as well as to provide assistance in breeding programs, in the development of postharvest technology and during the processing of scented rice cultivars. Moreover, this method has wide application in the analysis of rice to detect substitution. It can be effectively used in combination with methods for analysis of the DNA (Bradbury et al. 2005) to detect the mixture of fragrant and nonfragrant varieties. The two approaches together would allow analysis of the level of fragrance and the proportion of fragrant varieties in the sample (Mathure et al. 2011).

4.3.3.1 Analysis of Upland Scented Rice Cultivars

The analysis for upland scented rice cultivars was done with respect to volatiles other than 2AP. For this 15 volatiles were selected.

4.3.3.2 Development of Calibration Curve for Fifteen Other Aroma Volatiles

In addition to 2AP and seven volatiles, analytical grade GC standards of fifteen volatiles were used for identification of peak and quantification of volatiles. Calibration curves for volatiles were plotted (Fig. 4.10, Table 4.4).

4.3.3.3 Quantification of 2AP

The range of 2AP content among the cultivars under study varied from 0.038 (Phule radha) to 0.920 mg/kg (Kala bhat) (Table 4.5). Kala bhat recorded the highest content of 2AP (0.920 mg/kg), followed Kothmirsal (0.802 mg/kg). State-wise, among nonbasmati scented rice, the range of 2AP content varied from 0.038 (Phule radha) to 0.920 mg/kg (Kala bhat) in cultivars from Maharashtra, 0.079 (Sanna bili bhata) to 0.732 mg/kg (Kali kumud) in the cultivars from Karnataka, and 0.044 (Girija samba) to 0.787 mg/kg (Amritbhog) in cultivars from other states of India. Basmati 370 recorded the highest content of 2AP (0.434 mg/kg) in the basmati category. In nonscented rice, 2AP varied from 0.072 (Sonsali) to 0.125 mg/kg (Kolamb).

On the basis of 2AP content, Kali kumud and Basmati 370 were marked as extreme among nonbasmati scented cultivars from Karnataka and basmati categories respectively, while Kala bhat and Basmati 386 (0.230 mg/kg) were marked as outlier as depicted in a box plot (Fig. 4.11).

Among five rice categories, nonbasmati scented cultivars from Maharashtra recorded the highest average value for 2AP content (0.273 mg/kg), followed by nonbasmati scented cultivars from other states of India (0.260 mg/kg), nonbasmati scented cultivars from Karnataka (0.245 mg/kg), basmati (0.181 mg/kg), and nonscented (0.093 mg/kg) (Table 4.5). However, owing to the variation recorded within each category, the differences in the 2AP contents were nonsignificant among the rice categories as revealed by Duncan's multiple range test (Table 4.6).

The 2AP content in brown rice using steam distillation as extraction method is reported to vary as follows: 0.17 mg/kg (Buttery et al. 1983), 0.015–0.691 mg/kg (Widjaja et al. 1996), and 0.028 in Ghansal cultivar and 0.030 mg/kg in Pusa basmati (Nadaf et al. 2006). Yang et al. reported 2AP content of 4.9 μg/kg in black pigmented scented rice (2008a) and 1.69 μg/kg in Korean black rice (2008b) by Tenax trap method. Using solvent extraction, 2AP was reported in

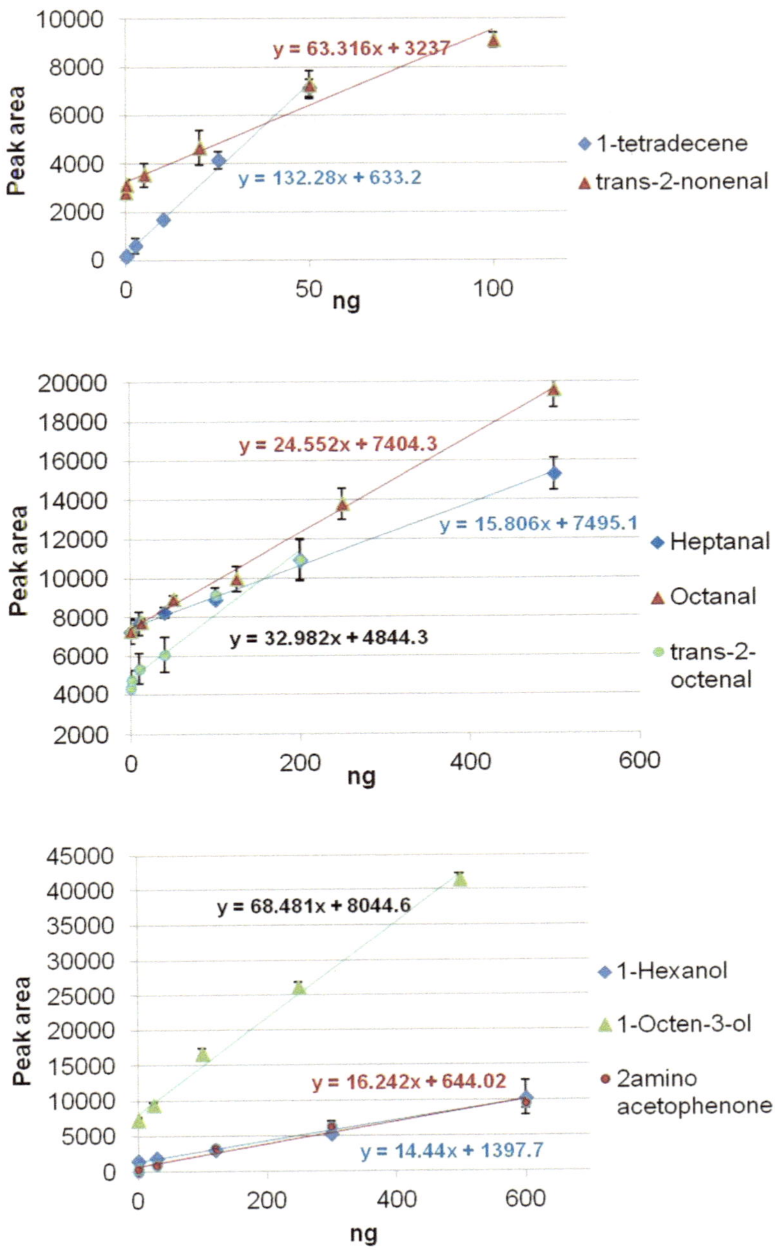

Fig. 4.10 Standard graphs for quantification of 15 volatiles

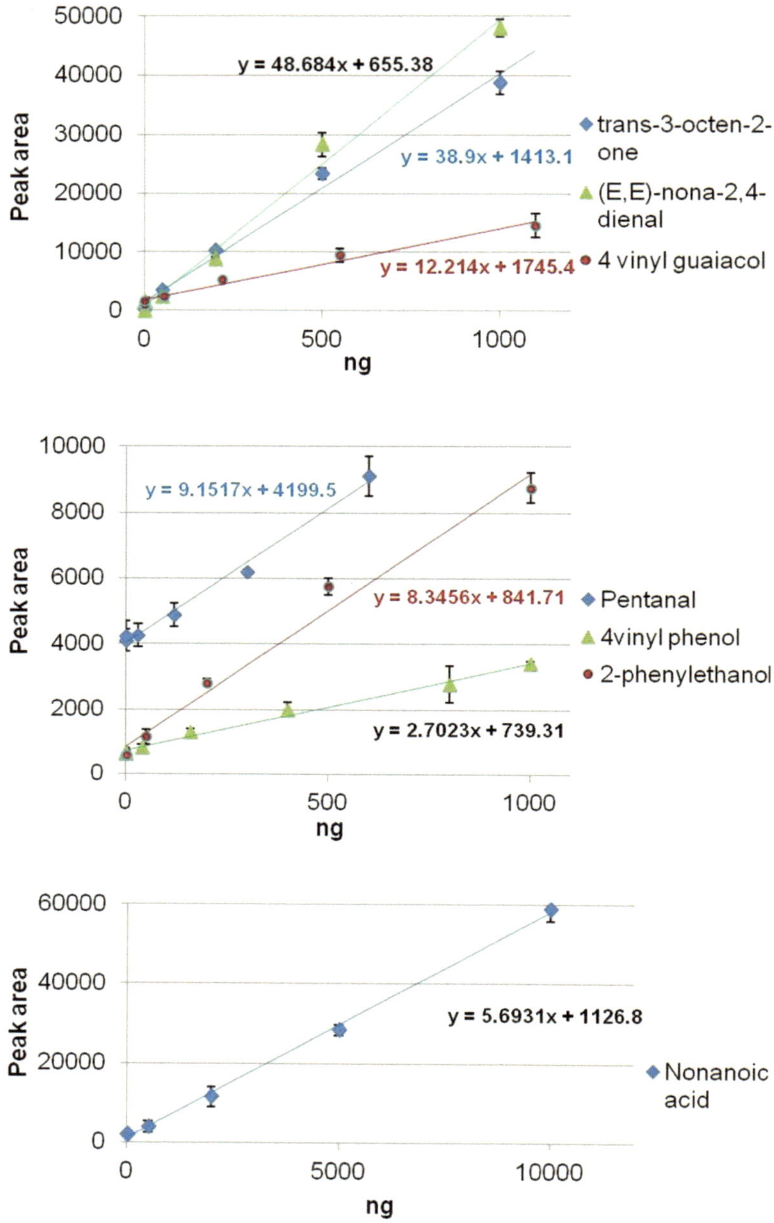

Fig. 4.10 (continued)

black glutinous rice (0.074–0.688 mg/kg) by Bounphanousay et al. (2008) and in basmati (0.119 mg/kg) by Bergman et al. (2000). Maraval et al. (2010) quantified 2AP using stable isotope dilution assay coupled with SPME-based analysis and reported 2AP content in scented rice (0.287–0.638 mg/kg) and nonscented rice (0.011–0.025 mg/kg).

Our earlier studies performed using marketed rice recorded 2AP content from 0.032 mg/kg in

Table 4.4 Retention time (RT), chemical class, validation range, and correlation coefficient (R^2) for 23 volatiles under study

RT (min)	Compound	Chemical class	Validation range (mg/kg)	R^2
1.88	Pentanal	Aliphatic aldehyde	0.001–0.6	0.946
2.72	Hexanal	Aliphatic aldehyde	0.125–5	0.991
4.13	Heptanal	Aliphatic aldehyde	0.01–0.5	0.996
5.99	Octanal	Aliphatic aldehyde	0.013–0.5	0.996
6.98	2-acetyl-1-pyrroline	N containing aromatic	0.020–1	0.987
7.5	1-hexanol	Aliphatic alcohol	0.001–0.6	0.994
8.1	Nonanal	Aliphatic aldehyde	0.015–0.6	0.989
8.26	*trans*-3-octen-2-one	Aliphatic ketone	0.001–1	0.989
8.66	*trans*-2-octenal	Aliphatic aldehyde	0.001–0.2	0.957
9.02	1-tetradecene	Aliphatic alkene	0.001–0.05	0.953
9.2	1-octen-3-ol	Aliphatic alcohol	0.001–0.121	0.993
10.11	Decanal	Aliphatic aldehyde	0.005–0.6	0.989
10.66	*trans*-2-nonenal	Aliphatic aldehyde	0.001–0.1	0.964
13.41	(*E*,*E*)-nona-2,4-dienal	Aliphatic aldehyde	0.001–1	0.991
15.36	Guaiacol	Aromatic	0.001–1	0.962
15.52	Benzyl alcohol	Aromatic	0.005–0.8	0.987
15.83	2-phenylethanol	Aromatic	0.001–1	0.982
18.05	4 vinyl guaiacol	Aromatic	0.001–1.1	0.979
18.15	Nonanoic acid	Aliphatic acid	0.020–10	0.998
18.21	2 amino acetophenone	N containing aromatic	0.001–0.6	0.978
19.34	4 vinyl phenol	Aromatic	0.001–1	0.978
19.83	Indole	N containing aromatic	0.002–0.5	0.981
20.58	Vanillin	Aromatic	0.20–0.8	0.987

Table partly adopted from Mathure et al. (2014)

Kolam type to 0.552 in Indrayani (Sect. 4.3.2.2). A marked increase in 2AP content of Ambemohar-157, Basmati 370, Basumati, Dubraj, Ghansal, Kali kumud, Kamavatya, and Kothmirsal in brown rice analyzed is observed in the present study as compared to the 2AP content recorded in milled rice (Sect. 4.3.2.2). However, few landraces, viz., Khadkya, Lal dodki, Manila, Raibhog, recorded a reduction in 2AP content than that reported in market rice. The difference in the quantity of 2AP due to variation in milling parameter is expected. However, the different environmental conditions during the growing season, the change in locality of cultivation, and postharvest conditions play a critical role resulting in variation in the quantity of 2AP. In the present study, 15 cultivars, viz., Ambemohar Ajra, Kothmirsal, Girga, Ghansal, Kala bhat, Kali kumud, Amritbhog, Dubraj, Dhanaprasad, Gatia, Kalajeera, Kalakrishna, Tulshimanjula, Ambemohar-157, and Indrayani, performed better in terms of 2AP content than basmati cultivars.

4.3.3.4 Quantification of Aromatic and Nitrogen Containing Aromatic Compounds

Banasphool-A recorded the highest content (0.933 mg/kg) of 4 vinyl phenol (Table 4.4). Minimum values for 4 vinyl phenol were recorded by Kalsal and Kondhekar chinoor (0.012 and 0.017 mg/kg respectively), and it was not detected in Badshahbhog and Parabhani chinoor. The content of 4 vinyl guaiacol varied from 0.008 (Chitak bhat and Kanakjeer) to 0.265 mg/kg (Ambemohar). Guaiacol was detected in 35 cultivars, and the content varied from 0.003 (Prabhatjeera) to 0.066 mg/kg (Ambemohar).

Table 4.5 Content of aromatic compounds in 91 rice (*Oryza sativa* L.) cultivars

Cultivar	2AP	Guaiacol	Benzyl alcohol	2-phenylethanol	4 vinyl guaiacol	2 amino acetophenone	4 vinyl phenol	Indole	Vanillin
Acharmati	0.077±0.012	nd	0.045±0.039	0.034±0.013	0.037±0.012	0.012±0.013	0.112±0.021	0.008±0.013	0.058±0.007
Ambemohar	0.099±0.086	0.066±0.058	0.028±0.025	nd	0.265±0.041	0.027±0.004	0.086±0.036	0.009±0.008	0.034±0.030
Ambemohar-157	0.728±0.013	nd	nd	0.041±0.004	0.023±0.004	0.006±0.010	0.111±0.026	nd	0.101±0.014
Ambemohar Ajra	0.434±0.012	nd	nd	0.038±0.005	0.021±0.003	0.014±0.001	0.072±0.008	nd	0.062±0.016
Ambemohar Pandhara	0.079±0.013	nd	nd	0.038±0.006	0.028±0.006	nd	0.060±0.058	0.020±0.006	0.064±0.001
Ambemohar-Tambda	0.108±0.026	nd	nd	0.046±0.008	0.030±0.007	nd	0.050±0.006	nd	0.091±0.022
Amritbhog	0.787±0.024	nd	nd	0.022±0.006	0.040±0.015	0.019±0.004	0.093±0.007	0.011±0.000	nd
Badsahbhog	0.087±0.075	nd	0.035±0.031	nd	0.018±0.019	0.024±0.021	nd	nd	nd
Bansaphool-A	0.107±0.010	nd	0.024±0.012	0.042±0.002	0.056±0.015	0.064±0.011	0.933±0.049	nd	0.063±0.028
Barke bhat	0.074±0.001	0.017±0.018	0.017±0.029	0.030±0.016	0.037±0.011	0.027±0.004	0.153±0.055	nd	0.145±0.008
Basmati 370	0.434±0.067	nd	0.015±0.001	0.027±0.012	0.035±0.009	0.016±0.005	0.036±0.062	nd	0.057±0.006
Basmati 376	0.108±0.022	0.031±0.010	nd	0.049±0.017	0.033±0.010	0.016±0.007	0.037±0.035	nd	0.041±0.004
Basmati 386	0.230±0.022	0.015±0.010	nd	0.049±0.012	0.045±0.027	0.021±0.005	0.096±0.005	nd	0.063±0.007
Basmati 6311	0.149±0.026	nd	0.069±0.006	0.073±0.016	0.045±0.006	0.025±0.005	0.058±0.050	0.007±0.012	0.030±0.026
Basumati	0.243±0.049	nd	0.020±0.002	0.529±0.696	0.023±0.029	nd	0.067±0.058	nd	nd
Bela blue	0.084±0.003	nd	nd	0.055±0.012	0.039±0.007	0.021±0.018	0.077±0.004	nd	0.040±0.004
Bhogavati	0.209±0.028	0.006±0.011	0.018±0.015	0.045±0.005	0.029±0.003	0.085±0.038	0.070±0.061	0.015±0.014	0.058±0.057
Bishnubhog	0.082±0.017	nd	0.010±0.009	0.046±0.002	0.023±0.005	0.002±0.004	0.065±0.009	nd	0.055±0.007
Champakali	0.367±0.013	nd	nd	0.033±0.005	0.034±0.003	0.006±0.005	0.084±0.010	0.008±0.001	nd
Chimansal	0.292±0.055	nd	nd	0.028±0.006	0.032±0.005	0.008±0.007	0.080±0.023	0.051±0.020	0.082±0.050
Chitak bhat	0.088±0.012	nd	nd	0.463±0.028	0.008±0.008	nd	0.095±0.025	nd	0.030±0.003
Dhanaprasad	0.459±0.063	0.005±0.005	0.010±0.017	0.040±0.024	0.025±0.010	0.017±0.014	0.062±0.015	nd	nd
Dubraj	0.540±0.052	0.010±0.003	nd	0.049±0.011	0.033±0.002	0.005±0.005	0.071±0.012	0.014±0.013	nd
Dubrajsena	0.394±0.053	nd	0.005±0.008	0.020±0.005	0.026±0.005	nd	0.075±0.007	nd	0.081±0.021
Dusara	0.159±0.006	nd	0.045±0.020	0.047±0.008	0.023±0.004	0.002±0.004	0.128±0.027	nd	0.041±0.020
Elaichi	0.227±0.052	0.018±0.002	nd	0.036±0.009	0.026±0.008	0.010±0.009	0.048±0.015	nd	0.046±0.009
Gandhesale	0.118±0.204	nd	0.009±0.015	0.096±0.166	0.011±0.018	0.003±0.005	0.027±0.046	nd	0.018±0.032
Gatia	0.450±0.011	0.023±0.002	nd	0.046±0.009	0.024±0.002	0.021±0.004	0.130±0.018	nd	0.061±0.011
Geerige sanna	0.300±0.025	nd	0.020±0.026	0.059±0.015	0.024±0.007	nd	0.090±0.031	nd	nd
Gham	0.099±0.008	nd	nd	0.034±0.001	0.030±0.009	nd	0.061±0.017	nd	0.059±0.004
Ghansal	0.462±0.029	nd	nd	0.028±0.010	0.031±0.006	0.004±0.007	0.108±0.035	nd	nd

Girga	0.473±0.017	nd	nd	0.031±0.007	0.032±0.011	0.014±0.007	0.081±0.008	nd	0.083±0.021
Girija samba	0.044±0.006	nd	0.047±0.016	0.026±0.032	0.020±0.018	0.017±0.015	0.052±0.052	nd	nd
Indrayani	0.441±0.018	nd	nd	0.032±0.006	0.033±0.008	0.103±0.012	0.056±0.007	nd	0.073±0.011
Jaya	0.102±0.031	nd	nd	0.037±0.008	0.045±0.008	nd	0.069±0.018	nd	0.061±0.009
Jeeraphool	0.225±0.030	0.008±0.000	0.036±0.009	0.029±0.010	0.028±0.005	nd	0.119±0.033	nd	0.008±0.013
Jeera-sona	0.148±0.008	nd	0.019±0.005	0.026±0.003	0.032±0.006	nd	0.092±0.017	nd	0.060±0.020
Jiri	0.072±0.030	nd	0.023±0.023	0.034±0.010	0.027±0.003	0.009±0.005	0.052±0.045	nd	0.074±0.025
Kagisali	0.249±0.014	nd	0.055±0.012	0.018±0.002	0.038±0.007	0.014±0.014	0.096±0.011	nd	0.062±0.072
Kala bhat	0.920±0.085	nd	0.019±0.005	0.039±0.005	0.047±0.014	0.015±0.002	0.076±0.006	nd	0.045±0.009
Kalakrishna	0.638±0.055	0.051±0.008	nd	0.022±0.014	0.023±0.003	0.014±0.006	0.105±0.030	nd	0.113±0.022
Kalajeera	0.727±0.004	nd	nd	0.034±0.012	0.031±0.007	0.007±0.008	0.037±0.032	0.017±0.009	nd
Kalanamak	0.142±0.123	nd	0.153±0.132	0.008±0.007	0.057±0.050	0.049±0.043	0.053±0.046	nd	0.022±0.021
Kali kumud	0.732±0.144	nd	nd	0.060±0.027	0.040±0.009	0.018±0.012	0.073±0.008	nd	0.049±0.014
Kaligajvili	0.179±0.011	0.006±0.005	0.035±0.004	0.017±0.004	0.028±0.004	0.024±0.007	0.094±0.012	nd	0.065±0.003
Kalsal	0.081±0.013	nd	nd	0.019±0.019	0.032±0.009	0.006±0.010	0.012±0.020	0.014±0.005	nd
Kamavatya	0.297±0.023	0.019±0.002	0.039±0.000	0.030±0.004	0.022±0.002	0.009±0.001	0.134±0.019	0.025±0.000	0.040±0.004
Kamod	0.383±0.022	0.007±0.006	0.006±0.010	0.026±0.005	0.031±0.004	0.008±0.008	0.089±0.030	0.014±0.003	0.031±0.015
Kanakjeer	0.430±0.009	nd	nd	0.031±0.005	0.008±0.009	0.015±0.005	0.057±0.049	0.018±0.005	0.038±0.003
Kate chinoor	0.385±0.014	nd	nd	0.036±0.015	0.036±0.011	0.007±0.006	0.071±0.010	0.033±0.005	0.095±0.004
Kernal local	0.148±0.004	0.006±0.011	0.093±0.004	0.059±0.013	0.088±0.085	0.258±0.038	0.105±0.020	nd	0.055±0.013
Khadkya	0.161±0.007	0.014±0.003	nd	0.026±0.006	0.061±0.009	0.010±0.004	0.076±0.009	nd	0.018±0.032
Kolamb	0.125±0.013	nd	0.035±0.006	0.034±0.003	0.051±0.003	0.265±0.185	0.103±0.013	nd	0.044±0.003
Kondhekar chinoor	0.120±0.031	nd	0.029±0.004	0.041±0.003	0.034±0.010	0.007±0.012	0.017±0.029	nd	0.032±0.004
Kothmbiri	0.227±0.029	0.006±0.005	0.018±0.004	0.039±0.011	0.031±0.014	0.023±0.007	0.119±0.025	nd	0.079±0.071
Kothmirsal	0.802±0.196	nd	nd	0.023±0.008	0.035±0.003	0.005±0.004	0.045±0.010	0.035±0.006	nd
Kumud	0.183±0.035	nd	0.025±0.005	0.031±0.007	0.044±0.025	0.052±0.013	0.092±0.027	nd	0.053±0.011
Lal bhat	0.315±0.020	nd	0.019±0.002	0.035±0.006	0.031±0.009	0.014±0.015	0.278±0.038	0.017±0.009	0.065±0.007
Lal dodki	0.222±0.030	nd	nd	0.043±0.006	0.034±0.009	0.013±0.010	0.298±0.047	0.025±0.006	0.078±0.009
Lalu	0.235±0.016	0.031±0.018	0.006±0.010	0.021±0.018	0.057±0.018	0.018±0.005	0.094±0.019	0.005±0.009	0.150±0.058
Makarand	0.235±0.012	nd	nd	0.017±0.009	0.035±0.006	0.092±0.023	0.071±0.007	nd	0.050±0.013
Manila	0.077±0.021	0.016±0.006	0.023±0.001	0.024±0.005	0.033±0.013	0.002±0.004	0.063±0.014	nd	0.058±0.010
Medhini sanna bhata	0.334±0.024	0.016±0.010	nd	0.032±0.002	0.020±0.003	nd	0.089±0.010	nd	0.043±0.010
Mugad sugandha	0.230±0.047	nd	0.020±0.003	0.039±0.011	0.028±0.004	0.048±0.021	0.069±0.004	nd	0.081±0.011

(continued)

Table 4.5 (continued)

Cultivar	2AP	Guaiacol	Benzyl alcohol	2-phenylethanol	4 vinyl guaiacol	2 amino acetophenone	4 vinyl phenol	Indole	Vanillin
Mysore sanna	0.129 ± 0.036	0.013 ± 0.005	nd	0.039 ± 0.000	0.036 ± 0.002	0.064 ± 0.059	0.051 ± 0.050	nd	0.049 ± 0.007
Pakhe bhat	0.063 ± 0.003	nd	nd	0.028 ± 0.010	0.024 ± 0.006	nd	0.060 ± 0.012	0.026 ± 0.008	0.068 ± 0.019
Pakistan basmati	0.125 ± 0.032	nd	0.026 ± 0.007	0.037 ± 0.004	0.042 ± 0.003	0.107 ± 0.037	0.077 ± 0.019	nd	0.167 ± 0.024
Parabhani chinoor	0.136 ± 0.003	nd	nd	0.041 ± 0.006	0.027 ± 0.007	0.009 ± 0.003	nd	0.003 ± 0.006	0.066 ± 0.011
Pawana	0.187 ± 0.008	0.010 ± 0.006	nd	0.041 ± 0.008	0.026 ± 0.012	0.055 ± 0.022	0.066 ± 0.039	nd	0.030 ± 0.034
Phule radha	0.038 ± 0.007	nd	0.029 ± 0.006	0.016 ± 0.004	0.043 ± 0.003	0.018 ± 0.006	0.053 ± 0.010	nd	0.062 ± 0.014
Pimpudibasa	0.418 ± 0.024	0.005 ± 0.008	0.043 ± 0.008	0.031 ± 0.019	0.040 ± 0.006	0.008 ± 0.008	0.101 ± 0.028	nd	0.061 ± 0.008
Prabhatjeera	0.122 ± 0.009	0.003 ± 0.005	0.013 ± 0.014	0.061 ± 0.009	0.033 ± 0.005	0.025 ± 0.008	0.062 ± 0.009	nd	0.043 ± 0.006
Pusa basmati	0.161 ± 0.014	0.024 ± 0.016	nd	0.074 ± 0.003	0.035 ± 0.009	0.024 ± 0.011	0.108 ± 0.056	0.007 ± 0.012	0.086 ± 0.006
Pusa basmati-1	0.153 ± 0.010	0.014 ± 0.005	nd	0.059 ± 0.008	0.032 ± 0.014	nd	0.060 ± 0.019	nd	0.033 ± 0.003
Pusa sugandha	0.141 ± 0.007	0.004 ± 0.007	0.034 ± 0.012	0.075 ± 0.027	0.044 ± 0.004	0.043 ± 0.005	0.132 ± 0.039	nd	0.021 ± 0.037
Raibhog	0.067 ± 0.020	0.053 ± 0.007	0.026 ± 0.010	0.036 ± 0.009	0.047 ± 0.009	0.005 ± 0.009	0.121 ± 0.048	0.009 ± 0.002	0.086 ± 0.035
Ratibhog	0.404 ± 0.034	nd	0.017 ± 0.002	0.036 ± 0.008	0.034 ± 0.004	0.042 ± 0.004	0.089 ± 0.009	nd	0.079 ± 0.009
RDN scented	0.129 ± 0.008	nd	0.035 ± 0.006	0.052 ± 0.010	0.027 ± 0.006	0.031 ± 0.009	0.058 ± 0.014	nd	0.048 ± 0.013
RDN local	0.050 ± 0.028	0.004 ± 0.007	0.040 ± 0.016	0.030 ± 0.010	0.035 ± 0.009	0.020 ± 0.020	0.062 ± 0.007	nd	0.065 ± 0.016
Sanna bili bhatta	0.079 ± 0.003	nd	0.061 ± 0.002	0.023 ± 0.013	0.041 ± 0.007	0.018 ± 0.003	0.110 ± 0.017	nd	nd
Shrabanmasi	0.048 ± 0.005	0.010 ± 0.010	0.014 ± 0.014	0.031 ± 0.018	0.028 ± 0.001	0.017 ± 0.012	0.083 ± 0.009	nd	0.047 ± 0.011
Shyamjeer	0.260 ± 0.030	nd	0.027 ± 0.005	0.011 ± 0.011	0.033 ± 0.009	0.011 ± 0.007	0.089 ± 0.020	nd	0.061 ± 0.020
Sonsali	0.072 ± 0.011	nd	0.055 ± 0.016	0.023 ± 0.005	0.029 ± 0.003	0.032 ± 0.006	0.283 ± 0.049	nd	0.060 ± 0.019
Super basmati	0.121 ± 0.009	nd	0.143 ± 0.012	0.027 ± 0.001	0.029 ± 0.006	0.105 ± 0.030	0.069 ± 0.018	nd	0.083 ± 0.020
Tamsal	0.071 ± 0.018	nd	nd	0.032 ± 0.005	0.036 ± 0.007	0.019 ± 0.003	0.187 ± 0.023	nd	0.050 ± 0.013
Tulshimanjula	0.453 ± 0.045	0.018 ± 0.009	nd	0.040 ± 0.008	0.037 ± 0.009	0.010 ± 0.013	0.037 ± 0.001	nd	0.131 ± 0.055
Tulsiamrit	0.149 ± 0.006	0.017 ± 0.001	0.021 ± 0.007	0.023 ± 0.006	0.021 ± 0.004	0.012 ± 0.011	0.097 ± 0.003	nd	0.413 ± 0.257
Tulsikanti	0.105 ± 0.007	0.005 ± 0.009	0.043 ± 0.007	0.028 ± 0.016	0.030 ± 0.004	0.015 ± 0.002	0.075 ± 0.015	0.005 ± 0.009	0.053 ± 0.012
Vasane sanna bhatta	0.171 ± 0.018	0.024 ± 0.041	0.029 ± 0.003	0.015 ± 0.013	0.027 ± 0.002	0.019 ± 0.018	0.107 ± 0.051	nd	0.024 ± 0.003
Velchi	0.224 ± 0.022	nd	nd	0.041 ± 0.006	0.035 ± 0.009	0.004 ± 0.007	0.059 ± 0.017	nd	nd
Velkat	0.069 ± 0.060	0.005 ± 0.008	0.027 ± 0.032	0.015 ± 0.019	0.024 ± 0.022	0.014 ± 0.014	0.066 ± 0.058	nd	0.090 ± 0.079
Min	**0.038**	**nd**	**nd**	**nd**	**0.008**	**nd**	**nd**	**nd**	**nd**
Max	**0.920**	**0.066**	**0.153**	**0.529**	**0.265**	**0.265**	**0.933**	**0.051**	**0.413**
Average	0.245	0.006	0.019	0.045	0.036	0.026	0.094	0.004	0.056
CV%	80.78	189.80	141.74	154.03	75.24	167.85	108.21	213.47	92.27

Values are in mg/kg, ±SD, *nd* not detected

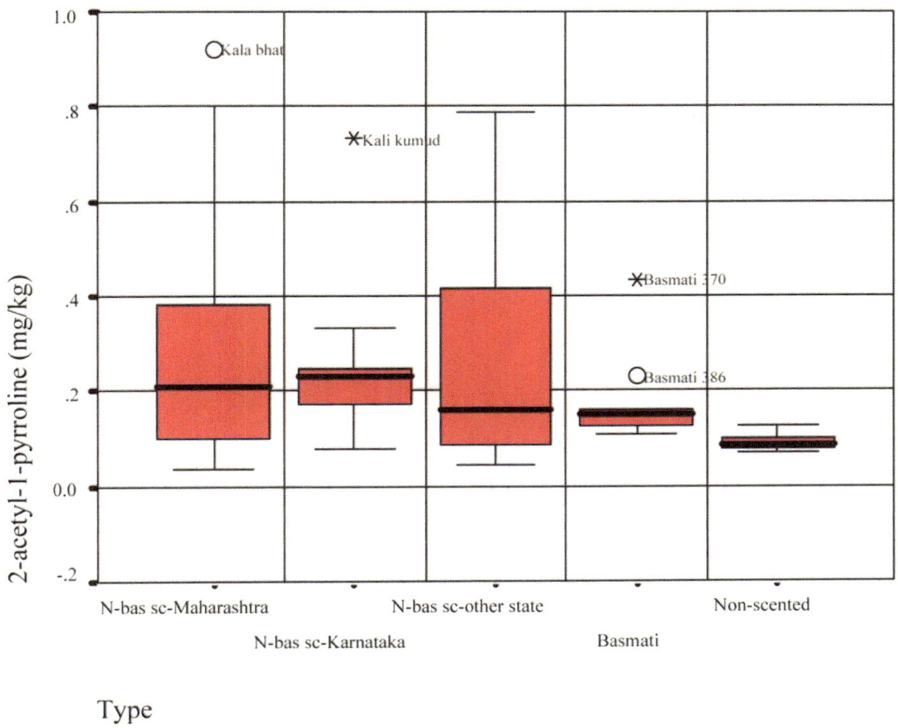

Fig. 4.11 Box plot based on 2-acetyl-1-pyrroline contents among the rice cultivars under study. Note the outliers (marked with a circle) and extremes (marked with an asterisk) among the rice categories (Source: Mathure et al. 2014). N-bas sc-Maharashtra: nonbasmati scented – Maharashtra; N-bas sc-Karnataka: nonbasmati scented – Karnataka; N-bas sc-other states: non basmati scented – other states; basmati: basmati; nonscented: nonscented

Vanillin was detected in 77 cultivars and varied from 0.008 mg/kg (Jeeraphool) to 0.413 mg/kg (Tulsiamrit). These include Basmati 386 (0.063 mg/kg), Pusa basmati (0.086 mg/kg), and Super basmati (0.083 mg/kg). Basumati recorded the highest content (0.529 mg/kg) of 2-phenylethanol, and lowest content of 0.008 mg/kg was recorded by Kalanamak. 2-phenylethanol was not detected in Ambemohar and Badshahbhog. Indole was detected in only 24 cultivars. Chimansal recorded the highest content (0.051 mg/kg) of indole, followed by Kothmirsal (0.035 mg/kg) and Kate chinoor (0.033 mg/kg). Among basmati types, indole was not detected, with the exception of Basmati 6311 and Pusa basmati. Benzyl alcohol was detected in 52 rice cultivars, and its content ranged from 0.005 mg/kg (Dubrajsena) to 0.153 mg/kg (Kalanamak). 2 amino acetophenone was detected in 60 cultivars. Kolamb recorded the highest content (0.265 mg/kg) of 2 amino acetophenone, followed by Kernal local (0.258 mg/kg).

Among aromatic and nitrogen containing aromatic compounds, 2-phenylethanol, benzyl alcohol, 2 amino acetophenone, and indole recorded significant variation among rice categories (Table 4.5). Nonscented cultivars recorded significantly lower average value (0.116 mg/kg) for 2-phenylethanol than scented rice cultivars. Higher content of benzyl alcohol was recorded in basmati cultivars than nonbasmati scented cultivars from Maharashtra. Basmati cultivars recorded significantly higher average value of 2 amino acetophenone (0.064 mg/kg) than nonbasmati scented rice cultivars. Indole recorded significantly higher content in nonbasmati scented rice from Maharashtra and was not detected in nonbasmati scented rice from Karnataka and

Table 4.6 Duncan's multivariate analysis of 23 volatiles among rice (*Oryza sativa* L.) categories

	Nonbasmati scented Maharashtra		Nonbasmati scented Karnataka		Nonbasmati scented other states		Basmati		Nonscented	
Compound	Average	CV%	Average	CV%	Average	CV%	Average	CV%	Average	CV%
Pentanal	0.076 a	71.82	0.078[a]	47.78	0.068 a	53.70	0.077 a	55.85	0.078 a	42.40
Hexanal	0.776 a	95.25	0.836 a	76.77	0.769 a	68.15	0.612 a	42.17	0.815 a	48.39
Heptanal	0.175 a	42.71	0.187 a	49.80	0.162 a	42.22	0.220 a	41.53	0.173 a	17.05
Octanal	0.094 a	41.73	0.117 a	36.45	0.107 a	50.49	0.211 b	58.96	0.110 a	31.22
2-acetyl-1-pyrroline	0.273 a	83.51	0.245 a	66.71	0.260 a	80.06	0.181 a	55.81	0.093 a	22.94
1-hexanol	0.201[ab]	52.37	0.189 ab	60.37	0.179 ab	50.60	0.231 b	34.05	0.117 a	61.40
Nonanal	0.215 a	40.49	0.263 ab	27.72	0.268 ab	44.29	0.361 b	59.60	0.244 a	33.67
trans-3-octen-2-one	0.010 a	142.54	0.020 a	119.84	0.012 a	150.83	0.009 a	94.86	0.013 a	117.34
trans-2-octenal	0.025 a	60.08	0.030 a	46.63	0.024 a	49.56	0.033 a	42.45	0.022 a	38.27
1-tetradecene	0.011 a	57.80	0.009 a	61.22	0.010 a	49.26	0.007 a	41.06	0.009 a	48.11
1-octen-3-ol	0.041 ab	38.89	0.029 ab	39.16	0.031 ab	37.36	0.045 a	90.87	0.027 b	27.63
Decanal	0.024 a	43.95	0.037 b	25.90	0.031 ab	31.51	0.050 c	10.75	0.032 ab	32.21
trans-2-nonenal	0.012 a	41.46	0.015 a	43.89	0.016 a	37.57	0.015 a	35.85	0.013 a	20.45
(*E,E*)-nona-2,4-dienal	0.004 a	105.76	0.003 a	145.02	0.004 a	110.20	0.006 a	94.82	0.004 a	64.90
Guaiacol	0.006 a	260.77	0.004 a	177.29	0.007 a	156.74	0.010 a	115.81	0.003 a	223.61
Benzyl alcohol	0.010 a	141.12	0.021 ab	96.34	0.021 ab	133.18	0.038 b	134.74	0.023 ab	104.67
2-phenylethanol	0.033 a	30.02	0.075 ab	184.28	0.033 a	46.26	0.050 a	34.89	0.116 b	167.38
4 vinyl guaiacol	0.040 a	104.91	0.030 a	32.24	0.032 a	34.45	0.043 a	42.21	0.033 a	50.80
Nonanoic acid	0.657 a	51.81	0.511 a	73.34	0.619 a	76.35	1.118 b	40.31	0.902 ab	96.22
2 amino acetophenone	0.017 a	136.57	0.027 ab	105.43	0.017 a	85.70	0.064 c	130.29	0.060 bc	192.62
4 vinyl phenol	0.086 a	74.91	0.080 a	29.16	0.108 a	140.37	0.072[a]	37.67	0.123 a	74.22
Indole	0.009 a	147.08	nd b	–	0.003 ab	209.59	0.002 ab	198.43	nd b	-
Vanillin	0.051 a	59.37	0.038 a	70.67	0.064 a	116.04	0.068 a	61.50	0.051 a	26.46

Values are in mg/kg. Values with the same letter in each rice category for each compound are not significantly different at $p=0.05$, nd: not detected

nonscented rice cultivars (source: Mathure et al. 2010).

Maraval et al. (2008) reported that 4 vinyl phenol quantities vary in a narrow range from 0.962 to 1.127 mg/kg using collidine as internal standard. These values are higher as compared to the values obtained in the present study (0.012–0.933 mg/kg). 4 vinyl guaiacol quantities reported in the present study are in agreement with the quantity reported by Buttery et al. (1988) and Widjaja et al. (1996). In comparison with our earlier study on milled rice (Sect. 4.3.2.2), brown rice recorded lower quantities of guaiacol. Earlier study (Sect. 4.3.2.2) recorded lower values for vanillin content in milled rice of Lal dodki, Raibhog, and Kali kumud as compared to the vanillin content obtained in the present study. Jezussek et al. (2002) have reported the presence of indole in Basmati 370 by aroma extract dilution analysis. However, in the present study, indole was detected only in Basmati 6311 and Pusa basmati among basmati cultivars. In the present study, the quantity of indole is lower than the values reported by Maraval et al. (2008) and in milled rice (Sect. 4.3.2.2).

4.3.3.5 Quantification of Aliphatic Aldehydes

Among nine aliphatic aldehydes, hexanal recorded highest average value of 0.768 mg/kg and 77.12 % coefficient of variation (Table 4.6). Hexanal was detected in all 91 cultivars and recorded values from 0.153 (Shyamjeer) to 4.304 mg/kg (Khadkya). Nonanal content ranged from 0.102 (Badshahbhog) to 0.875 mg/kg (Basmati 6311). Girija samba, Khadkya, and Pusa basmati recorded higher content of nonanal (0.584, 0.556, and 0.489 mg/kg respectively). Maximum octanal was recorded by Basmati 6311 (0.459 mg/kg), followed by Basmati 386 (0.344 mg/kg). Ambemohar, Badshahbhog, Kalsal, and Tamsal recorded octanal content lower than 0.05 mg/kg. Minimum *trans*-2-nonenal content of 0.007 mg/kg was recorded in Gandhesale, Girga, Konthekar chinoor, Makarand, Super basmati, and Tamsal. Medhini sanna bhata recorded 0.031 mg/kg of *trans*-2-nonenal. Maximum value of (*E*,*E*)-nona-2,4-dienal was recorded by Kernal local (0.018 mg/kg), followed by Ratibhog (0.016 mg/kg) and Kalanamak (0.015 mg/kg). Minimum amount of 0.001 mg/kg of (*E*,*E*)-nona-2,4-dienal was present in Gham, Ghansal, Kali kumud, and Pusa basmati. It was not detected in 27 rice cultivars. Heptanal quantity varied from 0.033 (Kalanamak) to 0.418 (Mugad sugandha) with an average value of 0.176 mg/kg.

Heptanal in basmati and nonscented rice varied from 0.084 (Pakistan basmati) to 0.300 (Basmati 376) and 0.129 (Sonsali) to 0.208 mg/kg (Chitak bhat). Lal dodki recorded lowest decanal content of 0.005 mg/kg, followed by Lal bhat (0.006 mg/kg), while Bhogavati recorded highest value of 0.060 mg/kg. Pentanal was not detected in Kalanamak, and Badshahbhog recorded 0.008 mg/kg quantity. Khadkya recorded highest pentanal content of 0.329 mg/kg, followed by Sanna bili bhata (0.180 mg/kg) and Basmati 386 (0.161 mg/kg). *Trans*-2-octenal content ranged from 0.003 (Kalanamak) to 0.067 mg/kg (Khadkya). It was not detected in Kalakrishna. In basmati, quantity of *trans*-2-octenal varied from 0.012 to 0.044 mg/kg.

Aliphatic aldehydes nonanal, octanal, and decanal recorded a significant variation in the rice cultivars belonging to different categories (Table 4.7). Basmati types recorded significantly higher average value for nonanal (0.361 mg/kg) than nonbasmati scented from Maharashtra (0.215 mg/kg) and nonscented cultivars (0.244 mg/kg). Basmati cultivars recorded significantly higher decanal content than nonbasmati scented and nonscented rice cultivars (Table 4.7). A significant variation in the quantity of decanal was observed in nonbasmati scented rice from Maharashtra (0.024 mg/kg) and from Karnataka (0.037 mg/kg).

Among nine aliphatic aldehydes, hexanal was detected in all cultivars. Bergman et al. (2000) reported 1.416 mg/kg hexanal in brown basmati rice. In cooked rice, hexanal content is reported to vary from 0.829 mg/kg in basmati to 2.038 mg/kg in nonscented Pelde rice (Widjaja et al. 1996). Yang et al. (2008b) reported hexanal content as 4.402 μg/kg relative concentration in black rice. The values obtained in the present study (0.153–

Table 4.7 Content of aliphatic aldehydes in 91 rice (*Oryza sativa* L.) cultivars

Name	Pentanal	Hexanal	Heptanal	Octanal	Nonanal	*Trans*-2-octenal	Decanal	*Trans*-2-nonenal	(E,E)-nona-2, 4-dienal
Acharmati	0.098±0.002	1.226±0.164	0.180±0.032	0.105±0.002	0.224±0.038	0.025±0.004	0.028±0.003	0.017±0.001	0.002±0.002
Ambemohar	0.015±0.014	0.452±0.392	0.047±0.041	0.036±0.032	0.217±0.188	0.014±0.004	0.013±0.012	0.012±0.001	nd
Ambemohar-157	0.080±0.005	0.790±0.059	0.160±0.022	0.076±0.012	0.144±0.004	0.012±0.002	0.021±0.002	0.009±0.000	0.004±0.001
Ambemohar Ajra	0.119±0.003	0.995±0.134	0.208±0.012	0.101±0.011	0.177±0.013	0.026±0.006	0.022±0.004	0.013±0.000	0.003±0.001
Ambemohar Pandhara	0.115±0.004	1.728±0.191	0.181±0.014	0.109±0.009	0.188±0.015	0.026±0.002	0.025±0.004	0.009±0.002	0.002±0.002
Ambemohar-Tambda	0.064±0.008	0.615±0.016	0.135±0.010	0.085±0.001	0.217±0.023	0.015±0.003	0.024±0.004	0.009±0.002	0.004±0.001
Amritbhog	0.113±0.007	0.905±0.062	0.200±0.032	0.070±0.009	0.120±0.017	0.012±0.005	0.022±0.001	0.009±0.001	0.009±0.003
Badsahbhog	0.008±0.007	0.166±0.144	0.042±0.037	0.035±0.030	0.102±0.092	0.003±0.005	0.016±0.014	0.011±0.010	0.006±0.005
Bansaphool-A	0.151±0.017	1.231±0.192	0.211±0.017	0.262±0.020	0.444±0.069	0.039±0.036	0.051±0.013	0.023±0.003	0.010±0.002
Barke bhat	0.099±0.010	0.974±0.117	0.156±0.008	0.126±0.016	0.385±0.070	0.025±0.001	0.046±0.009	0.020±0.000	nd
Basmati 370	0.085±0.007	0.502±0.045	0.256±0.070	0.166±0.070	0.311±0.023	0.038±0.006	0.043±0.004	0.011±0.006	nd
Basmati 376	0.048±0.004	0.419±0.021	0.300±0.005	0.195±0.007	0.189±0.026	0.040±0.006	0.053±0.003	0.016±0.001	0.003±0.004
Basmati 386	0.161±0.048	0.869±0.233	0.319±0.050	0.344±0.092	0.387±0.045	0.044±0.011	0.043±0.004	0.014±0.003	0.009±0.003
Basmati 6311	0.118±0.004	1.191±0.267	0.234±0.022	0.459±0.029	0.875±0.352	0.044±0.019	0.054±0.012	0.024±0.002	nd
Basumati	0.090±0.078	1.836±0.580	0.206±0.189	0.162±0.149	0.304±0.050	0.037±0.032	0.039±0.008	0.009±0.008	0.004±0.003
Bela blue	0.081±0.012	0.554±0.130	0.118±0.020	0.121±0.029	0.162±0.026	0.023±0.007	0.024±0.002	0.013±0.001	0.005±0.002
Bhogavati	0.076±0.007	0.467±0.071	0.345±0.038	0.145±0.012	0.253±0.021	0.045±0.006	0.060±0.007	0.014±0.002	0.011±0.005
Bishnubhog	0.060±0.006	0.633±0.045	0.140±0.006	0.083±0.008	0.203±0.012	0.015±0.006	0.030±0.005	0.016±0.005	0.004±0.007
Champakali	0.108±0.029	1.035±0.252	0.203±0.040	0.129±0.022	0.257±0.028	0.023±0.007	0.016±0.004	0.011±0.002	0.003±0.001
Chimansal	0.104±0.003	0.784±0.082	0.141±0.020	0.080±0.010	0.160±0.009	0.025±0.003	0.017±0.004	0.009±0.001	0.002±0.002
Chitak bhat	0.099±0.003	1.355±0.054	0.208±0.007	0.148±0.019	0.317±0.009	0.032±0.002	0.024±0.002	0.010±0.001	0.008±0.001
Dhanaprasad	0.056±0.004	0.615±0.158	0.250±0.059	0.162±0.035	0.375±0.118	0.048±0.018	0.034±0.004	0.021±0.003	0.004±0.002
Dubraj	0.046±0.011	0.627±0.088	0.166±0.021	0.091±0.010	0.295±0.149	0.033±0.025	0.037±0.016	0.017±0.005	nd
Dubrajsena	0.075±0.004	0.682±0.108	0.170±0.018	0.100±0.014	0.188±0.028	0.026±0.003	0.034±0.004	0.012±0.001	0.005±0.003
Dusara	0.039±0.006	0.300±0.003	0.068±0.010	0.064±0.008	0.147±0.006	0.026±0.008	0.037±0.007	0.013±0.003	0.004±0.004
Elaichi	0.075±0.010	1.011±0.382	0.212±0.041	0.145±0.031	0.297±0.043	0.038±0.014	0.052±0.017	0.018±0.003	nd
Gandhesale	0.048±0.083	0.577±1.000	0.091±0.158	0.050±0.087	0.156±0.270	0.013±0.023	0.017±0.029	0.007±0.012	0.002±0.004

Gatia	0.046±0.002	0.564±0.018	0.169±0.015	0.066±0.016	0.134±0.006	0.023±0.003	0.020±0.005	0.010±0.002	0.006±0.000
Geerige sanna	0.092±0.023	0.765±0.342	0.215±0.054	0.181±0.072	0.348±0.044	0.038±0.007	0.048±0.006	0.023±0.003	0.005±0.005
Gham	0.117±0.010	1.354±0.161	0.165±0.020	0.097±0.014	0.201±0.009	0.062±0.003	0.028±0.004	0.010±0.002	0.001±0.002
Ghansal	0.072±0.006	0.603±0.101	0.218±0.048	0.088±0.020	0.145±0.031	0.025±0.002	0.026±0.004	0.012±0.001	0.001±0.001
Girga	0.056±0.009	0.470±0.051	0.148±0.023	0.062±0.002	0.168±0.006	0.019±0.005	0.023±0.002	0.007±0.006	0.003±0.003
Girija samba	0.059±0.005	0.781±0.019	0.069±0.008	0.086±0.010	0.584±0.102	0.028+0.007	0.027±0.007	0.026±0.008	nd
Indrayani	0.053±0.006	0.540±0.110	0.178±0.002	0.125±0.012	0.217±0.016	0.025±0.005	0.036±0.002	0.009±0.001	0.013±0.005
Jaya	0.045±0.035	0.457±0.181	0.166±0.038	0.083±0.018	0.139±0.031	0.015±0.002	0.029±0.004	0.011±0.001	0.005±0.001
Jeeraphool	0.098±0.006	0.865±0.039	0.268±0.020	0.120±0.015	0.368±0.057	0.040±0.006	0.026±0.003	0.020±0.002	nd
Jeera-sona	0.030±0.002	0.304±0.044	0.085±0.007	0.062±0.002	0.318±0.061	0.023±0.006	0.027±0.003	0.023±0.002	nd
Jiri	0.034±0.001	0.734±0.288	0.138±0.033	0.076±0.039	0.134±0.070	0.043±0.042	0.021±0.001	0.013±0.006	nd
Kagisali	0.078±0.005	0.776±0.103	0.142±0.008	0.086±0.011	0.182±0.011	0.022±0.004	0.037±0.004	0.011±0.002	nd
Kala bhat	0.034±0.004	0.379±0.103	0.158±0.008	0.113±0.009	0.286±0.018	0.016±0.005	0.031±0.008	0.014±0.002	0.003±0.001
Kalakrishna	0.063±0.008	1.282±0.328	0.171±0.021	0.068±0.007	0.166±0.017	nd	0.024±0.003	nd	nd
Kalajeera	0.048±0.014	0.487±0.216	0.203±0.032	0.102±0.022	0.158±0.016	0.017±0.004	0.032±0.004	0.010±0.002	nd
Kalanamak	nd	3.033±4.331	0.033±0.029	0.063±0.055	0.253±0.219	0.003±0.005	0.022±0.019	0.002±0.004	0.015±0.013
Kali kumud	0.068±0.019	0.406±0.038	0.204±0.041	0.115±0.021	0.261±0.029	0.031±0.008	0.037±0.011	0.014±0.002	0.001±0.002
Kaligajvili	0.058±0.014	0.383±0.022	0.091±0.016	0.095±0.009	0.250±0.042	0.021±0.003	0.029±0.006	0.013±0.005	nd
Kalsal	0.039±0.008	0.380±0.141	0.099±0.009	0.043±0.007	0.167±0.046	0.007±0.002	0.020±0.004	0.009±0.002	nd
Kamavatya	0.045±0.004	1.148±0.032	0.145±0.015	0.188±0.020	0.320±0.012	0.034±0.002	0.017±0.000	0.012±0.000	nd
Kamod	0.051±0.001	0.383±0.081	0.135±0.005	0.092±0.024	0.139±0.023	0.023±0.003	0.021±0.001	0.009±0.004	nd
Kanakjeer	0.113±0.019	0.942±0.083	0.159±0.029	0.068±0.005	0.138±0.018	0.017±0.002	0.023±0.002	0.013±0.003	0.005±0.001
Kate chinoor	0.064±0.005	0.503±0.044	0.184±0.003	0.088±0.005	0.209±0.035	0.010±0.001	0.026±0.001	0.011±0.001	nd
Kernal local	0.026±0.005	0.456±0.096	0.084±0.015	0.087±0.008	0.219±0.042	0.013±0.002	0.053±0.009	0.010±0.006	0.018±0.007
Khadkya	0.329±0.011	4.304±0.579	0.366±0.066	0.214±0.028	0.556±0.106	0.067±0.011	0.046±0.007	0.023±0.002	nd
Kolamb	0.105±0.008	1.101±0.290	0.189±0.033	0.138±0.008	0.337±0.046	0.028±0.010	0.050±0.001	0.015±0.001	0.005±0.007
Kondhekar chinoor	0.106±0.002	0.616±0.213	0.393±0.015	0.096±0.004	0.187±0.016	0.028±0.001	0.023±0.005	0.007±0.007	0.002±0.001
Kothmbiri	0.030±0.004	0.338±0.018	0.129±0.007	0.109±0.009	0.446±0.037	0.018±0.001	0.034±0.005	0.021±0.002	nd
Kothmirsal	0.099±0.011	0.869±0.173	0.201±0.016	0.102±0.003	0.191±0.024	0.023±0.001	0.027±0.003	0.013±0.001	0.004±0.001
Kumud	0.071±0.016	0.526±0.051	0.157±0.017	0.077±0.017	0.225±0.027	0.025±0.001	0.043±0.006	0.018±0.005	nd

(continued)

Table 4.7 (continued)

Name	Pentanal	Hexanal	Heptanal	Octanal	Nonanal	*Trans*-2-octenal	Decanal	*Trans*-2-nonenal	(E,E)-nona-2, 4-dienal
Lal bhat	0.046 ± 0.001	0.411 ± 0.021	0.108 ± 0.021	0.055 ± 0.009	0.148 ± 0.012	0.003 ± 0.005	0.006 ± 0.001	nd	0.006 ± 0.002
Lal dodki	0.051 ± 0.003	0.369 ± 0.012	0.118 ± 0.012	0.053 ± 0.007	0.134 ± 0.018	0.015 ± 0.001	0.005 ± 0.000	0.009 ± 0.003	0.010 ± 0.005
Lalu	0.052 ± 0.008	0.616 ± 0.032	0.213 ± 0.039	0.134 ± 0.006	0.370 ± 0.135	0.020 ± 0.007	0.029 ± 0.007	0.022 ± 0.003	0.008 ± 0.003
Makarand	0.049 ± 0.009	0.350 ± 0.042	0.233 ± 0.021	0.085 ± 0.005	0.143 ± 0.005	0.021 ± 0.002	0.029 ± 0.002	0.007 ± 0.002	0.012 ± 0.000
Manila	0.100 ± 0.011	0.659 ± 0.090	0.171 ± 0.032	0.112 ± 0.021	0.224 ± 0.045	0.024 ± 0.005	0.029 ± 0.011	0.016 ± 0.003	nd
Medhini sanna bhata	0.104 ± 0.015	0.809 ± 0.054	0.247 ± 0.020	0.154 ± 0.005	0.374 ± 0.030	0.031 ± 0.004	0.042 ± 0.002	0.031 ± 0.012	nd
Mugad sugandha	0.066 ± 0.005	0.680 ± 0.354	0.418 ± 0.023	0.179 ± 0.019	0.248 ± 0.032	0.055 ± 0.009	0.042 ± 0.008	0.013 ± 0.001	0.013 ± 0.007
Mysore sanna	0.091 ± 0.007	0.850 ± 0.158	0.159 ± 0.018	0.124 ± 0.027	0.312 ± 0.010	0.027 ± 0.007	0.044 ± 0.001	0.018 ± 0.003	0.003 ± 0.004
Pakhe bhat	0.044 ± 0.000	0.275 ± 0.024	0.120 ± 0.020	0.050 ± 0.003	0.115 ± 0.012	0.023 ± 0.001	0.022 ± 0.004	0.009 ± 0.000	0.002 ± 0.003
Pakistan basmati	0.062 ± 0.008	0.402 ± 0.019	0.142 ± 0.008	0.091 ± 0.012	0.211 ± 0.007	0.018 ± 0.003	0.043 ± 0.002	0.013 ± 0.001	0.007 ± 0.006
Parabhani chinoor	0.070 ± 0.005	0.483 ± 0.049	0.152 ± 0.053	0.084 ± 0.009	0.199 ± 0.006	0.017 ± 0.004	0.026 ± 0.002	0.015 ± 0.006	0.003 ± 0.003
Pawana	0.097 ± 0.001	1.055 ± 0.067	0.175 ± 0.009	0.119 ± 0.001	0.337 ± 0.019	0.025 ± 0.005	0.028 ± 0.001	0.013 ± 0.004	0.012 ± 0.003
Phule radha	0.042 ± 0.010	0.282 ± 0.040	0.174 ± 0.015	0.059 ± 0.008	0.194 ± 0.033	0.032 ± 0.003	0.023 ± 0.001	0.020 ± 0.005	nd
Pimpudibasa	0.124 ± 0.027	1.423 ± 0.390	0.242 ± 0.027	0.149 ± 0.018	0.371 ± 0.051	0.039 ± 0.007	0.052 ± 0.015	0.023 ± 0.003	nd
Prabhatjeera	0.135 ± 0.020	1.172 ± 0.086	0.300 ± 0.020	0.204 ± 0.048	0.351 ± 0.020	0.049 ± 0.001	0.043 ± 0.007	0.018 ± 0.000	0.005 ± 0.004
Pusa basmati	0.077 ± 0.005	0.597 ± 0.197	0.275 ± 0.014	0.245 ± 0.048	0.489 ± 0.145	0.041 ± 0.010	0.056 ± 0.009	0.020 ± 0.004	0.001 ± 0.002
Pusa basmati-1	0.080 ± 0.004	0.509 ± 0.090	0.282 ± 0.013	0.204 ± 0.026	0.330 ± 0.045	0.043 ± 0.014	0.049 ± 0.004	0.016 ± 0.001	0.007 ± 0.002
Pusa sugandha	0.113 ± 0.032	0.782 ± 0.187	0.232 ± 0.069	0.258 ± 0.013	0.432 ± 0.122	0.038 ± 0.006	0.041 ± 0.019	0.018 ± 0.007	0.012 ± 0.002
Raibhog	0.039 ± 0.005	0.193 ± 0.037	0.107 ± 0.011	0.064 ± 0.006	0.170 ± 0.009	0.027 ± 0.002	0.020 ± 0.006	0.010 ± 0.003	0.011 ± 0.004
Ratibhog	0.042 ± 0.003	0.362 ± 0.014	0.234 ± 0.008	0.113 ± 0.011	0.329 ± 0.053	0.024 ± 0.003	0.047 ± 0.010	0.017 ± 0.004	0.016 ± 0.013
RDN scented	0.058 ± 0.008	0.649 ± 0.114	0.127 ± 0.015	0.104 ± 0.056	0.348 ± 0.054	0.034 ± 0.008	0.032 ± 0.007	0.023 ± 0.009	0.006 ± 0.003
RDN local	0.042 ± 0.003	0.350 ± 0.047	0.208 ± 0.028	0.065 ± 0.017	0.209 ± 0.027	0.032 ± 0.004	0.031 ± 0.005	0.020 ± 0.007	0.002 ± 0.004
Sanna bili bhatta	0.180 ± 0.011	2.540 ± 0.055	0.219 ± 0.022	0.132 ± 0.012	0.280 ± 0.074	0.058 ± 0.008	0.048 ± 0.006	0.016 ± 0.003	nd
Shrabanmasi	0.064 ± 0.011	0.592 ± 0.124	0.134 ± 0.024	0.079 ± 0.012	0.267 ± 0.085	0.020 ± 0.006	0.032 ± 0.002	0.014 ± 0.005	0.002 ± 0.003

Shyamjeer	0.020±0.002	0.153±0.015	0.062±0.014	0.056±0.006	0.290±0.047	0.013±0.001	0.023±0.002	0.014±0.002	0.003±0.002
Sonsali	0.038±0.010	0.504±0.139	0.129±0.007	0.068±0.005	0.202±0.024	0.012±0.003	0.030±0.006	0.012±0.003	0.004±0.004
Super basmati	0.031±0.002	0.559±0.138	0.088±0.003	0.105±0.015	0.242±0.060	0.012±0.011	0.053±0.004	0.007±0.005	0.013±0.001
Tamsal	0.062±0.020	0.315±0.008	0.114±0.014	0.050±0.005	0.122±0.005	0.005±0.006	0.014±0.001	0.007±0.001	0.007±0.001
Tulshimanjula	0.065±0.008	0.654±0.118	0.225±0.030	0.112±0.009	0.199±0.018	0.028±0.005	0.035±0.007	0.016±0.005	0.015±0.002
Tulsiamrit	0.040±0.012	0.379±0.038	0.071±0.017	0.060±0.008	0.183±0.018	0.023±0.004	0.024±0.004	0.019±0.003	0.004±0.001
Tulsikanti	0.092±0.006	0.761±0.088	0.195±0.001	0.118±0.002	0.261±0.015	0.023±0.006	0.028±0.004	0.016±0.002	0.002±0.002
Vasane sanna bhatta	0.025±0.004	0.370±0.020	0.046±0.001	0.078±0.012	0.339±0.073	0.012±0.007	0.025±0.006	0.019±0.008	nd
Velchi	0.075±0.009	0.797±0.133	0.201±0.019	0.121±0.003	0.269±0.010	0.007±0.007	0.016±0.001	0.009±0.002	0.002±0.002
Velkat	0.060±0.052	0.697±0.611	0.129±0.112	0.072±0.063	0.169±0.147	0.012±0.016	0.019±0.017	0.010±0.009	nd
Min	**nd**	**0.153**	**0.33**	**0.035**	**0.102**	**nd**	**0.005**	**nd**	**nd**
Max	**0.329**	**4.304**	**0.418**	**0.459**	**0.875**	**0.067**	**0.060**	**0.031**	**0.018**
Average	0.074	0.768	0.176	0.114	0.257	0.026	0.032	0.014	0.004
CV%	59.04	77.12	42.97	57.41	46.70	51.85	38.65	40.46	108.69

Values are in mg/kg, ±SD, *nd* not detected

4.304 mg/kg) are in agreement with the values reported by Widjaja et al. (1996). Hexanal content in brown rice is lower than the values obtained in milled rice in earlier study (Sect. 4.3.2.2). In India, rice is stored for about a year before its sale to obtain preferred cooking qualities. This storage, along with the processes like milling, can contribute in the increase of hexanal content. Such increase in hexanal content with storage and milling is reported by researchers (Wongpornchai et al. 2003; Champagne 2008; Tananuwong and Lertsiri 2010). Widjaja et al. (1996) recorded values for nonanal (0.125–0.429 mg/kg); octanal (0.058–0.105 mg/kg); *trans*-2-nonenal (0.024–0.067 mg/kg); (*E*,*E*)-nona-2,4-dienal (0.006–0.021 mg/kg); heptanal 0.111 (basmati) to 0.132 mg/kg (nonscented rice); and decanal. The quantities reported in the present study are in agreement with these reports. However, quantity of *trans*-2-octenal in basmati recorded in the present study is lower than 0.055 mg/kg as reported by Widjaja et al. (1996). In the present study, basmati recorded significantly higher nonanal content than nonscented cultivars. Similar findings are reported by Yang et al. (2008a).

4.3.3.6 Quantification of Aliphatic Alcohols, Ketone, Alkene, and Acid

Quantity of 1-octen-3-ol varied from 0.012 (Dusara) to 0.121 mg/kg (Super basmati) (Table 4.8). Highest value for nonanoic acid was reported by Sonsali (2.360 mg/kg), while it was not detected in Chimansal and Gandhesale.

On the other hand, 1-tetradecene content ranged from 0.002 (Gandhesale, Sanna bili bhata, and Velkat) to 0.033 mg/kg (Kala bhat). Khadkya recorded highest content of 0.596 mg/kg of 1-hexanol. It was not detected in Chitak bhat. *Trans*-3-octen-2-one was detected in 62 cultivars from 0.001 (Gandhesale, Khadkya, Velkat, and Pusa basmati-1) to 0.090 mg/kg (Banasphool-A).

Aliphatic alcohols, viz., 1-octen-3-ol and 1-hexanol and nonanoic acid, recorded variation with respect to the rice category (Table 4.5). Quantity of 1-octen-3-ol recorded higher value in basmati (0.045 mg/kg) than nonscented (0.027 mg/kg) rice cultivars. Basmati recorded significantly higher content of nonanoic acid in comparison with nonbasmati scented and nonscented rice. In present the study, 1-hexanol content in basmati cultivars was significantly higher than nonscented, while content in nonbasmati scented was not significantly different than either of these categories (Table 4.5) (Mathure et al. 2014).

Reports on 1-octen-3-ol in basmati vary from 0.046 mg/kg (Widjaja et al. 1996) to 0.360 mg/kg (Tava and Bocchi 1999). Our results are in agreement with the reported values. Nonanoic acid was not detected in Chimansal and Gandhesale, while highest content of 2.360 mg/kg was recorded by Sonsali. A study by Maraval et al. (2008) reported the presence of nonanoic acid in traces. Content of 1-tetradecene ranged from 0.002 (Gandhesale, Sanna bili bhata, and Velkat) to 0.033 mg/kg (Kala bhat). Ajarayasiri and Chaiseri (2008) reported lower content of 1-tetradecene in black glutinous rice.

Trans-3-octen-2-one was detected in 62 cultivars from 0.001 (Gandhesale, Khadkya, Velkat, and Pusa basmati-1) to 0.090 mg/kg (Banasphool-A). These values are in agreement with the report by Maraval et al. (2008). Aliphatic alcohol 1-octen-3-ol recorded higher content in basmati than nonscented rice cultivars. However, Widjaja et al. (1996) reported higher content of 1-octen-3-ol in nonscented rice than basmati. This inconsistency can be attributed to the difference in growing conditions and postharvest conditions of basmati and nonscented rice analyzed by Widjaja et al. (1996). Basmati recorded significantly higher content of nonanoic acid in comparison with nonbasmati scented and nonscented rice. Petrov et al. (1996), Maraval et al. (2008) and Yang et al. (2008a) reported that 1-hexanol can discriminate between scented and nonscented cultivars. In the present study, 1-hexanol content in basmati cultivars was significantly higher than nonscented, while content in nonbasmati scented was not significantly different than either of these categories (Table 4.5).

Table 4.8 Content of aliphatic acid, alcohol, alkene, and ketone in 91 rice (*Oryza sativa* L.) cultivars

Name	1-hexanol	*Trans*-3-octen-2-one	1-tetradecene	1-octen-3-ol	Nonanoic acid
Acharmati	0.204±0.039	nd	0.006±0.002	0.039±0.003	0.371±0.048
Ambemohar	0.084±0.075	0.017±0.001	0.004±0.002	0.030±0.009	0.207±0.044
Ambemohar-157	0.150±0.015	0.009±0.002	0.023±0.002	0.045±0.004	0.789±0.111
Ambemohar Ajra	0.301±0.048	0.018±0.003	0.014±0.002	0.039±0.005	0.563±0.056
Ambemohar Panchara	0.230±0.074	nd	0.007±0.001	0.043±0.001	0.759±0.103
Ambemohar-Tambda	0.143±0.020	0.006±0.001	0.009±0.001	0.055±0.003	0.615±0.154
Amritbhog	0.239±0.037	nd	0.023±0.003	0.036±0.003	0.748±0.173
Badsahbhog	0.070±0.061	nd	0.006±0.005	0.014±0.012	0.461±0.400
Bansaphool-A	0.375±0.029	0.090±0.034	0.008±0.002	0.046±0.002	2.284±0.856
Barke bhat	0.122±0.008	0.013±0.002	0.008±0.001	0.028±0.003	0.444±0.076
Basmati 370	0.237±0.040	nd	0.014±0.002	0.025±0.005	1.240±0.652
Basmati 376	0.271±0.029	0.018±0.001	0.004±0.000	0.016±0.002	0.797±0.093
Basmati 386	0.304±0.044	nd	0.007±0.000	0.023±0.005	0.838±0.138
Basmati 6311	0.315±0.082	0.006±0.011	0.007±0.002	0.031±0.005	0.877±0.177
Basumati	0.239±0.220	0.007±0.006	0.005±0.005	0.031±0.028	0.042±0.037
Bela blue	0.148±0.014	0.024±0.008	0.007±0.001	0.021±0.002	0.510±0.080
Bhogavati	0.313±0.055	0.040±0.002	0.007±0.000	0.022±0.001	1.891±0.314
Bishnubhog	0.229±0.021	0.013±0.006	0.006±0.001	0.027±0.003	0.537±0.076
Champakali	0.165±0.019	nd	0.021±0.005	0.053±0.002	0.693±0.036
Chimansal	0.286±0.026	nd	0.008±0.002	0.036±0.005	nd
Chitak bhat	nd	0.037±0.010	0.016±0.002	0.015±0.001	0.081±0.007
Dhanaprasad	0.172±0.011	0.008±0.002	0.017±0.002	0.043±0.008	0.381±0.071
Dubraj	0.138±0.021	0.005±0.001	0.009±0.000	0.039±0.003	0.364±0.095
Dubrajsena	0.191±0.022	0.021±0.005	0.010±0.002	0.029±0.001	0.745±0.195
Dusara	0.132±0.004	0.021±0.003	0.008±0.002	0.012±0.001	0.258±0.019
Elaichi	0.228±0.052	0.021±0.003	0.010±0.001	0.034±0.012	0.715±0.424
Gandhesale	0.078±0.134	0.001±0.001	0.002±0.004	0.015±0.026	nd
Gatia	0.258±0.024	0.048±0.005	0.017±0.001	0.030±0.004	0.654±0.086
Geerige sanna	0.221±0.047	0.077±0.034	0.011±0.003	0.043±0.007	0.583±0.171
Gham	0.298±0.048	nd	0.003±0.000	0.040±0.002	0.348±0.002
Ghansal	0.192±0.018	0.008±0.000	0.016±0.001	0.032±0.006	0.686±0.076
Girga	0.123±0.047	nd	0.014±0.001	0.039±0.004	0.421±0.045
Girija samba	0.123±0.029	nd	0.007±0.000	0.024±0.001	0.315±0.284
Indrayani	0.205±0.038	0.035±0.005	0.015±0.001	0.027±0.002	0.936±0.103
Jaya	0.179±0.042	0.009±0.002	0.010±0.001	0.035±0.001	0.749±0.116
Jeeraphool	0.235±0.038	nd	0.008±0.001	0.034±0.003	0.403±0.019
Jeera-sona	0.120±0.011	nd	0.006±0.002	0.021±0.001	0.267±0.030
Jiri	0.325±0.017	0.008±0.002	0.006±0.000	0.038±0.010	0.527±0.047
Kagisali	0.155±0.027	0.017±0.006	0.007±0.003	0.022±0.002	0.806±0.072
Kala bhat	0.092±0.009	nd	0.033±0.001	0.040±0.003	0.844±0.075
Kalakrishna	0.245±0.028	nd	0.016±0.002	0.029±0.001	1.010±0.150
Kalajeera	0.145±0.066	0.006±0.002	0.020±0.001	0.035±0.003	0.416±0.077
Kalanamak	0.050±0.044	nd	0.003±0.002	0.072±0.064	0.895±0.778
Kali kumud	0.156±0.045	0.011±0.001	0.023±0.001	0.043±0.008	0.461±0.088

(continued)

Table 4.8 (continued)

Name	1-hexanol	*Trans*-3-octen-2-one	1-tetradecene	1-octen-3-ol	Nonanoic acid
Kaligajvili	0.114±0.026	0.014±0.012	0.007±0.002	0.013±0.001	0.275±0.074
Kalsal	0.135±0.011	nd	0.005±0.001	0.039±0.010	0.436±0.101
Kamavatya	0.133±0.042	0.060±0.005	0.008±0.001	0.053±0.005	0.835±0.017
Kamod	0.185±0.031	0.007±0.001	0.018±0.001	0.030±0.001	0.590±0.053
Kanakjeer	0.298±0.034	nd	0.018±0.000	0.028±0.002	1.031±0.150
Kate chinoor	0.138±0.018	nd	0.016±0.002	0.037±0.004	0.520±0.106
Kernal local	0.117±0.009	0.018±0.001	0.005±0.001	0.111±0.006	1.376±0.179
Khadkya	0.596±0.014	0.001±0.002	0.010±0.004	0.066±0.008	0.594±0.047
Kolamb	0.139±0.013	nd	0.005±0.001	0.030±0.008	0.856±0.400
Kondhekar chinoor	0.274±0.036	0.002±0.002	0.009±0.001	0.047±0.003	0.679±0.083
Kothmbiri	0.085±0.010	0.031±0.002	0.008±0.002	0.024±0.002	0.469±0.040
Kothmirsal	0.340±0.006	nd	0.020±0.004	0.045±0.004	0.462±0.075
Kumud	0.158±0.003	0.015±0.003	0.011±0.000	0.027±0.001	0.320±0.039
Lal bhat	0.120±0.026	nd	0.015±0.003	0.047±0.003	0.571±0.146
Lal dodki	0.128±0.012	0.006±0.000	0.013±0.001	0.080±0.013	1.064±0.549
Lalu	0.098±0.020	nd	0.009±0.002	0.031±0.003	0.587±0.188
Makarand	0.009±0.002	0.009±0.002	0.011±0.001	0.018±0.001	0.855±0.367
Manila	0.166±0.063	0.021±0.003	0.007±0.001	0.027±0.003	0.465±0.041
Medhini sanna bhata	0.289±0.075	nd	0.009±0.002	0.035±0.005	0.384±0.060
Mugad sugandha	0.394±0.058	0.068±0.013	0.006±0.001	0.024±0.006	1.418±0.863
Mysore sanna	0.153±0.021	0.017±0.001	0.011±0.001	0.037±0.006	0.313±0.039
Pakhe bhat	0.047±0.013	0.006±0.001	0.007±0.002	0.043±0.006	0.512±0.039
Pakistan basmati	0.138±0.050	0.020±0.004	0.008±0.002	0.025±0.001	0.632±0.255
Parabhani chinoor	0.112±0.007	0.010±0.001	0.009±0.001	0.049±0.003	0.625±0.099
Pawana	0.232±0.024	0.020±0.003	0.008±0.002	0.031±0.001	1.233±0.247
Phule radha	0.200±0.098	nd	0.007±0.001	0.014±0.002	0.297±0.063
Pimpudibasa	0.274±0.024	0.003±0.005	0.012±0.000	0.045±0.005	0.352±0.067
Prabhatjeera	0.337±0.026	0.008±0.000	0.008±0.001	0.032±0.001	0.801±0.249
Pusa basmati	0.228±0.056	0.007±0.001	0.007±0.002	0.027±0.002	2.101±0.516
Pusa basmati-1	0.317±0.046	0.001±0.003	0.006±0.001	0.023±0.003	0.876±0.258
Pusa sugandha	0.350±0.071	0.037±0.004	0.009±0.001	0.040±0.011	2.234±0.479
Raibhog	0.099±0.041	0.009±0.002	0.011±0.004	0.082±0.006	0.622±0.092
Ratibhog	0.099±0.009	0.020±0.001	0.011±0.002	0.031±0.002	0.278±0.042
RDN scented	0.215±0.045	0.022±0.028	0.003±0.000	0.022±0.007	0.413±0.269
RDN local	0.180±0.028	0.015±0.001	0.006±0.001	0.015±0.004	0.498±0.208
Sanna bili bhatta	0.388±0.042	0.025±0.005	0.002±0.000	0.049±0.001	0.659±0.196
Shrabanmasi	0.116±0.028	0.028±0.007	0.009±0.002	0.021±0.003	0.639±0.046
Shyamjeer	0.039±0.016	nd	0.008±0.002	0.014±0.002	0.330±0.069
Sonsali	0.101±0.029	nd	0.006±0.000	0.027±0.013	2.360±0.822
Super basmati	0.151±0.030	0.006±0.006	0.005±0.000	0.121±0.020	1.328±0.147
Tamsal	0.093±0.010	0.004±0.002	0.008±0.001	0.027±0.003	1.021±0.410
Tulshimanjula	0.177±0.035	0.007±0.008	0.009±0.001	0.041±0.006	0.432±0.064
Tulsiamrit	0.123±0.035	nd	0.007±0.000	0.020±0.002	0.333±0.040
Tulsikanti	0.286±0.011	nd	0.006±0.001	0.030±0.001	0.388±0.033

(continued)

Table 4.8 (continued)

Name	1-hexanol	*Trans*-3-octen-2-one	1-tetradecene	1-octen-3-ol	Nonanoic acid
Vasane sanna bhatta	0.106±0.019	0.003±0.006	0.005±0.000	0.024±0.006	0.527±0.110
Velchi	0.159±0.012	0.004±0.001	0.006±0.001	0.057±0.010	0.637±0.101
Velkat	0.145±0.127	0.001±0.002	0.002±0.002	0.021±0.018	0.265±0.229
Min	**nd**	**nd**	**0.002**	**0.012**	**nd**
Max	**0.596**	**0.090**	**0.033**	**0.121**	**2.360**
Average	0.190	0.012	0.010	0.036	0.682
CV%	51.89	140.29	55.45	51.81	68.19

Values are in mg/kg, ±SD, *nd* not detected

Table 4.9 Correlation among 23 volatile compounds studied in 91 rice (*Oryza sativa* L.) cultivars

Name of compound	RT	1.88	2.72	4.13	5.99	6.98	7.50	8.10	8.26	8.66	9.02	9.20	10.11	10.66	13.41	15.36	15.52	15.83	18.05	18.15	18.21	19.34	19.83
Pentanal	1.88																						
Hexanal	2.72	0.703																					
Heptanal	4.13	0.590	0.273																				
Octanal	5.99	0.524	0.310	0.603																			
2-acetyl-1-pyrroline	6.98	-0.003	-0.060	0.160	-0.038																		
1-Hexanol	7.50	0.743	0.514	0.660	0.530	0.051																	
Nonanal	8.10	0.393	0.344	0.303	0.762	-0.139	0.296																
trans-3-octen-2-one	8.26	0.059	0.004	0.192	0.286	-0.085	0.181	0.139															
trans-2-octenal	8.66	0.596	0.402	0.666	0.622	-0.128	0.682	0.486	0.294														
1-tetradecene	9.02	0.045	-0.108	0.131	-0.065	0.838	-0.040	-0.156	-0.056	-0.164													
1-Octen-3-ol	9.20	0.083	0.264	-0.068	-0.007	0.066	0.066	-0.012	-0.032	-0.073	0.080												
Decanal	10.11	0.324	0.191	0.487	0.616	-0.059	0.397	0.534	0.301	0.561	-0.171	0.033											
trans-2-nonenal	10.66	0.279	0.064	0.241	0.429	-0.167	0.289	0.673	0.174	0.509	-0.174	-0.225	0.496										
(E,E)-nona-2,4-dienal	13.41	-0.132	-0.045	0.074	0.066	-0.001	-0.062	-0.099	0.279	-0.083	-0.011	0.340	0.154	-0.231									
Guaiacol	15.36	-0.094	-0.022	-0.022	0.031	-0.044	-0.014	0.067	0.016	-0.002	-0.091	0.012	0.000	0.066	-0.062								
Benzyl alcohol	15.52	-0.208	0.204	-0.295	0.021	-0.300	-0.116	0.192	0.017	-0.091	-0.380	0.396	0.211	-0.008	0.221	-0.098							
2-phenylethanol	15.83	0.107	0.194	0.130	0.201	-0.056	-0.038	0.131	0.125	0.165	0.011	-0.078	0.084	-0.080	0.090	-0.105	-0.083						
4 vinyl guaicol	18.05	-0.053	0.044	-0.118	-0.014	-0.088	-0.051	0.063	0.019	-0.049	-0.116	0.155	-0.004	0.004	0.039	0.496	0.162	-0.151					
Nonanoic acid	18.15	0.099	0.021	0.272	0.361	-0.040	0.268	0.160	0.335	0.101	-0.002	0.221	0.354	-0.046	0.408	-0.003	0.174	-0.141	0.020				
2amino acetophenone	18.21	-0.069	-0.026	-0.039	0.059	-0.123	-0.087	0.071	0.146	-0.057	-0.154	0.275	0.401	-0.056	0.484	-0.068	0.372	-0.072	0.218	0.347			
4vinyl phenol	19.34	0.144	0.043	-0.038	0.169	-0.071	0.106	0.126	0.442	0.014	0.013	0.121	0.060	0.097	0.168	-0.013	0.054	-0.012	0.089	0.458	0.105		
Indole	19.83	-0.002	-0.049	-0.073	-0.108	0.214	0.030	-0.200	-0.127	-0.137	0.194	0.170	-0.308	-0.238	-0.106	-0.030	-0.193	-0.085	0.018	-0.066	-0.150	0.008	
Vanillin	20.58	-0.160	-0.198	-0.140	-0.128	-0.092	-0.148	-0.131	-0.041	-0.120	-0.077	-0.064	0.000	0.026	0.116	0.206	-0.030	-0.138	-0.036	0.008	0.065	0.082	-0.049

Correlation is significant at the 0.01 level (2-tailed).

Correlation is significant at the 0.05 level (2-tailed).

4.3.3.7 Correlation among Headspace Volatiles in Rice Cultivars

Correlation analysis of 23 compounds revealed 69 significant correlations (Table 4.9). 2-acetyl-1-pyrroline correlated positively with 1-tetradecene (r=0.838) and negatively with benzyl alcohol (r=–0.3) at 0.01 significance level. It also exhibited a positive correlation with indole (r=0.214) at 0.05 significance level.

Hexanal was positively correlated with pentanal (r=0.703), heptanal (r=0.273), octanal (r=0.31), 1-hexanol (r=0.514), nonanal (r=0.344), *trans*-2-octenal (r=0.402) at 0.01 significance and with 1-octen-3-ol (r=0.264) at 0.05 significance level. Pentanal, heptanal, octanal, nonanal, *trans*-2-octenal among aliphatic aldehydes and 1-hexanol correlated positively with all aliphatic aldehydes except (*E,E*) nona-2,4 dienal. Pentanal correlated negatively with benzyl alcohol (r=–0.208) at 0.05 significance. Heptanal correlated with benzyl alcohol (r=–0.295) and nonanoic acid (0.272). Octanal exhibited positive correlation with *trans*-3-octen-2-one (r=0.286) and nonanoic acid (r=0.361) at 0.01 significance. *Trans*-2-octenal correlated with *trans*-3-octen-2-one (r=0.294) (Mathure et al. 2014).

Positive correlation is obtained in decanal with pentanal ($r=0.324$), heptanal ($r=0.487$), octanal ($r=0.616$), 1-hexanol ($r=0.397$), nonanal ($r=0.534$), *trans*-3-octene-2-one ($r=0.301$), *trans*-2-octenal ($r=0.561$), *trans*-2-nonenal ($r=0.496$), benzyl alcohol ($r=0.211$), nonanoic acid ($r=0.354$), 2 amino acetophenone ($r=0.401$). Indole and decanal correlated negatively ($r=0.308$) at 0.01 significance level. *Trans*-2-nonanal correlated negatively with 1-octen-3-ol ($r=-0.225$), (*E*,*E*)-nona-2,4 dienal ($r=-0.231$), and indole ($r=-0.238$) and positively with pentanal ($r=0.279$), heptanal ($r=0.241$), octanal ($r=0.429$), nonanal ($r=0.673$), *trans*-2-octenal ($r=0.509$), and 1-hexanol ($r=0.289$). Positive correlation of (*E*,*E*)-nona-2,4 dienal with aliphatic ketone (*trans*-3-octen-2-one) ($r=0.279$) and 1-octen-2-ol ($r=0.34$) at 0.01 significance level and with benzyl alcohol ($r=0.221$) was observed. 1-tetradecene was negatively related with benzyl alcohol ($r=-0.38$). 1-octen-3-ol correlated positively with nonanoic acid ($r=0.221$) and benzyl alcohol ($r=0.396$). 2 amino acetophenone exhibited significant positive relation with 1-octen-3-ol ($r=0.275$), nonanoic acid ($r=0.347$), benzyl alcohol ($r=0.372$), and 4 vinyl guaiacol ($r=0.218$). Guaiacol and 4 vinyl guaiacol correlated positively ($r=0.496$) at 0.01 significance level. 4 vinyl phenol correlated positively with aliphatic ketone, i.e., *trans*-3-octene-2-one ($r=0.442$) and aliphatic acid (nonanoic acid ($r=0.458$)). Nonanoic acid further exhibited correlation with 1-hexanol ($r=0.268$) and *trans*-3-octene-2-one ($r=0.335$). 2 phenyl ethanol and vanillin did not exhibit correlation with any other compound (Mathure et al. 2014).

Hexanal, pentanal, heptanal, octanal, nonanal, *trans*-2-octenal, decanal, *trans*-2-nonenal, (*E*,*E*)-2,4-nonadienal, 1-octen-3-ol, and 1-hexanol are the products derived from either oxidation or degradation of lipids (Lam and Proctor 2003, Mansoor and Proctor 2004, Mildner-Szkudlarz et al. 2003, Frankel 2005). Hence, they are related with each other resulting in significant correlation.

4.3.3.8 Principle Component Analysis

Principle component analysis of 23 headspace volatiles yielded eight principle components (PCs) with eigenvalue higher than 1 explaining 75.12 % cumulative variance. PC1, PC2, PC3, PC4, PC5, PC6, PC7, and PC8 explained 22.552, 12.667, 9.882, 7.593, 6.893, 5.645, 4.961, and 4.913 % variance respectively. Contribution of each compound in PC is presented in Table 4.10.

As seen from Table 4.9, majority of aliphatic aldehydes, viz., pentanal, hexanal, heptanal, octanal, nonanal, *trans*-2-octenal, decanal, and *trans*-2-nonenal, along with 1-hexanol, displayed significant loadings (value higher than 0.4) for first component. Benzyl alcohol and 2 amino acetophenone exhibited highest loadings of 0.69 each in second component, followed by (*E*,*E*)-nona-2,4-dienal and nonanoic acid. 2-acetyl-1-pyrroline and 1-tetradecene recorded negative projection on second component. Indole and vanillin did not exhibit significant contribution in any of the eight components extracted.

Among the volatiles studied, ten compounds, viz., hexanal, 2-acetyl-1-pyrroline, 1-tetradecene, 1-octe-3-ol, *trans*-2-nonenal, (*E*,*E*)-nona-2,4-dienal, benzyl alcohol, 2-phenylethanol, nonanoic acid, and 4 vinyl phenol, displayed significant loadings in more than one principle component, thus suggesting involvement of multiple factors either as a result of synthesis, accumulation, and degradation in plant system or due to certain uncontrollable causes during postharvest and analysis.

4.3.3.9 Contribution of Volatile Compounds to Rice Aroma

Fifteen compounds exceed the odor active value (OAV) higher than 1 in one or more cultivars (Table 4.11).

2AP recorded maximum OAV of 9195.85. Odor of 2AP is described as pandan, cooked rice, sweet, pleasant, and popcorn like (Table 4.10). Aliphatic aldehydes, viz., hexanal (956.53), nonanal (874.72), octanal (655.73), *trans*-2-nonenal (393.33), (*E*,*E*)-nona-2,4-dienal (261.84), heptanal

Table 4.10 Contribution of 23 volatiles in principle components extracted by principle component analysis

	Component[a]							
Compound	1	2	3	4	5	6	7	8
Pentanal	**0.740**	−0.332	0.124	0.316	0.053	−0.178	−0.098	−0.065
Hexanal	**0.538**	−0.090	0.055	**0.658**	−0.017	−0.126	0.097	−0.032
Heptanal	**0.724**	−0.308	0.243	−0.132	−0.008	0.069	−0.061	−0.340
Octanal	**0.843**	0.008	0.037	−0.114	0.021	0.102	0.116	0.094
2-acetyl-1-pyrroline	−0.122	**−0.469**	**0.622**	−0.149	0.154	**0.426**	0.075	0.140
1-hexanol	**0.758**	−0.252	0.171	0.176	0.133	−0.183	−0.252	−0.147
Nonanal	**0.720**	0.089	−0.270	−0.012	0.065	0.258	0.140	0.379
trans-3-octen-2-one	0.367	0.293	0.194	**−0.432**	−0.048	−0.371	0.270	0.088
trans-2-octenal	**0.833**	−0.180	−0.129	−0.014	−0.015	−0.043	−0.026	−0.162
1-tetradecene	−0.158	**−0.485**	**0.646**	−0.197	0.112	0.313	0.128	0.191
1-octen-3-ol	0.026	0.367	**0.469**	**0.543**	0.129	0.095	−0.035	0.076
Decanal	**0.734**	0.280	−0.021	−0.197	−0.069	0.358	−0.075	−0.072
trans-2-nonenal	**0.573**	−0.042	**−0.468**	−0.285	0.131	0.186	−0.118	0.326
(*E*,*E*)-nona-2,4-dienal	0.036	**0.570**	**0.497**	−0.104	−0.180	0.040	0.099	−0.326
Guaiacol	−0.010	0.097	−0.206	−0.075	**0.766**	0.016	0.302	−0.288
Benzyl alcohol	0.036	**0.690**	−0.112	**0.407**	−0.149	0.196	−0.062	0.240
2-phenylethanol	0.167	−0.120	−0.011	0.030	**−0.475**	−0.092	**0.714**	−0.171
4 vinyl guaiacol	−0.019	0.320	−0.087	0.193	**0.694**	0.078	0.348	−0.056
Nonanoic acid	0.357	**0.467**	**0.483**	−0.215	0.072	−0.157	−0.187	0.049
2 amino acetophenone	0.105	**0.690**	0.230	0.008	−0.069	0.308	−0.068	−0.145
4 vinyl phenol	0.206	0.323	0.290	−0.243	0.155	**−0.556**	0.064	**0.431**
Indole	−0.220	−0.283	0.314	0.225	0.239	−0.235	−0.113	0.161
Vanillin	−0.168	0.190	−0.118	−0.335	0.206	−0.079	−0.336	−0.336
% variance	22.552	12.677	9.882	7.593	6.893	5.645	4.961	4.913
% cumulative variance	22.552	35.229	45.112	52.705	59.599	65.243	70.204	75.118

[a]Positive or negative contribution of value higher than 0.4 is highlighted as bold

(139.40), decanal (29.89), pentanal (27.39), and *trans*-2-octenal (22.20), were major contributors to scent (Mathure et al. 2014). Pentanal was not detected in Kalanamak, and Badshahbhog recorded 0.008 mg/kg quantity, which is below the odor threshold value; except for these cultivars, pentanal will contribute to scent. As the odor threshold of *trans*-2-octenal is 3 μg/kg, it will contribute to the scent of all cultivars, except Kalakrishna.

Aromatic compounds 4 vinyl phenol, 4 vinyl guaiacol, guaiacol, and vanillin recorded OAVs of 93.34, 88.19, 22.08, and 7.12 respectively. Forty-three cultivars recorded vanillin higher than its odor value. Basmati 370 recorded vanillin content lower than its odor threshold value. The present study indicates that indole does not contribute to rice aroma. Aliphatic alcohol-1-octen-3-ol contributed to scent with OAV of 121.06. Since odor thresholds of 2 amino

Table 4.11 Odor description, reported odor thresholds of 23 volatiles

RT	Volatile Name	Odor description[a]	Odor threshold (μg/kg)[b]
1.88	Pentanal	Nutty, sweet	12a
2.72	Hexanal	Green, grasslike, herbal	4.5b
4.13	Heptanal	Floral	3a
5.99	Octanal	Slightly fruity, citrus like	0.7a
6.98	2-acetyl-1-pyrroline	Pandan, cooked rice, sweet, pleasant, popcorn	0.1a
7.50	1-hexanol	Vegetal, green	2500a
8.10	Nonanal	Flour, sticky rice, fatty, floral, fruity	1a
8.26	*trans*-3-octen-2-one	Green, fruity, rose	–
8.66	*trans*-2-octenal	Green, fatty, nutty, cooked flour	3a
9.02	1-tetradecene	–	60c
9.20	1-octen-3-ol	Straw, earthy, raw mushroom	1a
10.11	Decanal	Sweet, waxy, floral	2a
10.66	*trans*-2-nonenal	Fatty, woody, beany cucumber	0.08a
13.41	(*E,E*)-nona-2,4-dienal	Sweet, fatty, cooked flour	0.07a
15.36	Guaiacol	Black rice like, smoke	3a
15.52	Benzyl alcohol	Slightly sweet	10,000a
15.83	2-phenylethanol	Sweet, floral, fruity, flowery	1100a
18.05	4 vinyl guaiacol	Sweet, spicy, clove like, smoky	3a
18.15	Nonanoic acid	Animal, cheese	3000a
18.21	2 amino acetophenone	Medicinal, phenolic	–
19.34	4 vinyl phenol	Phenolic, medicinal	10a
19.83	Indole	Sweet, burnt, floral	140a
20.58	Vanillin	Vanilla	58b

[a]1. Ajarayasiri and Chaiseri (2008), 2. Widjaja et al. (1996), 3. Maraval et al. (2008), 4. Yang et al. (2008a), 5. Jezusek et al. (2002)

[b]a. Buttery et al. (1988), b. Buttery et al. (1999), c Ajarayasiri and Chaiseri (2008)

acetophenone and *trans*-3-octen-2-one were not available, their contribution to scent was not determined (Mathure et al. 2014).

Our study confirms the role of 2-acetyl-1-pyrroline as a key odorant in rice as reported by other researchers (Ajarayasiri and Chaiseri 2008; Yang et al. 2008a, b; Jezussek et al. 2002; Maraval et al. 2008). In the present study, on the basis of OAV, aliphatic aldehydes, viz., hexanal, nonanal, octanal, *trans*-2-nonenal, (*E,E*)-nona-2,4-dienal, heptanal, decanal, pentanal, and *trans*-2-octenal, were the major contributors in rice aroma. This finding is supported by Yang et al. (2008b). They considered aldehydes as important contributors to black rice due to their lower odor thresholds. Hexanal is a lipid oxidation product (Zhou et al. 2002) and gives off flavor to rice (Wongpornchai et al. 2003; Lam and Proctor 2003, Sect. 4.3.2.2). Odor of nonanal is described as flour, sticky rice, fatty, floral, and fruity (Ajarayasiri and Chaiseri 2008; Widjaja et al. 1996). It is also a lipid oxidation product (Zhou el al. 2002) and contributes to rice aroma (Sect. 4.3.2.2). *Trans*-2-nonenal contributed as off flavor (Lam and Proctor 2003) to the scent of all rice (Yang et al. 2008a). (*E,E*)-nona-2,4-dienal was identified as key odorant by Ajarayasiri and Chaiseri (2008) in black glutinous rice and Yang et al. (2008b) in black rice. Heptanal is reported as key odorant attributing floral tone in rice scent (Yang et al. 2008a). Decanal is reported as one of the contributors in rice aroma by Maraval et al. (2008). Pentanal was not detected in Kalanamak, and Badshahbhog recorded 0.008 mg/kg quantity, which is below the odor threshold value. Hence, except for these cultivars, pentanal will contribute to the scent of rice. Yang et al. (2008a) detected pentanal in basmati and described its odor as nutty and sweet. As the odor threshold of *trans*-2-octenal is 3 μg/kg, it will contribute to the scent of all cultivars, except Kalakrishna (Mathure et al. 2014).

Aromatic compounds 4 vinyl phenol, 4 vinyl guaiacol, guaiacol, and vanillin also contribute to aroma. Buttery et al. (1988) reported 4 vinyl phenol as one of the major contributors in rice odor. 4 vinyl guaiacol is reported as an important odorant in black rice by Jezussek et al. (2002) and Maraval et al. (2008). Guaiacol attributes to the characteristic aroma of black rice (Yang et al. 2008b) and Indian scented rice cultivars (Sect. 4.3.2.2).

Forty-three cultivars recorded vanillin higher than its odor value. Basmati 370 recorded vanillin content lower than its odor threshold value. Vanillin content increases upon cooking (Zeng et al. 2009) and is an important odorant in rice (Jezussek et al. 2002; Maraval et al. 2008, Sect. 4.3.2.2). Indole is reported to contribute to the aroma of white glutinous rice (Ajarayasiri and Chaiseri 2008) and cooked black rice (Yang et al. 2008b). However, in an earlier study on marketed rice (Sect. 4.3.2.2) and in the present study, it did not contribute to rice aroma (Mathure et al. 2014).

Jezussek et al. (2002) reported 2 amino acetophenone as an important odorant and described its odor as medicinal, phenolic. Aliphatic alcohol-1-octen-3-ol contributed to scent with higher OAV. This is in agreement with Ajarayasiri and Chaiseri (2008), Yang et al. (2008b), and Zeng et al. (2009)).

Though the rice cultivars were subjected to the same agroecological and postharvest conditions, a great degree of variation was recorded in 15 aroma active volatiles. Among these, nonanal, octanal, decanal, and 1-octen-3-ol could separate nonscented cultivars from basmati cultivars. Owing to the variation in nonbasmati scented cultivars from Maharashtra, Karnataka, and other parts of India, they either mingle with basmati or nonscented cultivars, thus indicating the need for studying scented rice cultivars beyond basmati cultivars. As a trend, basmati and other hybrid cultivars are cultivated so as to gain higher market price. But as cultivars like basmati are native to the northern region of India, when cultivated in higher average day temperature record reduction in 2AP content as compared to locally adapted cultivars is evident from this study. The local cultivars can create a better opportunity in scented rice research owing to their adaptability, variation in the composition of volatiles, and sustenance to produce higher quality of aroma and thus the consumer's preference.

4.4 Cultivar Specific Volatile Profile

With respect to odor-active volatiles, cultivar specific volatile profiles were recorded by the cultivars under study. The graphical representation of profiles is depicted in Figs. 4.12, 4.13, 4.14, 4.15, 4.16, 4.17, 4.18, 4.19, 4.20, and 4.21. Each volatile demonstrated distinct variation in its content among the cultivars. Thus, these odor-active volatiles can be taken to distinguish the cultivars

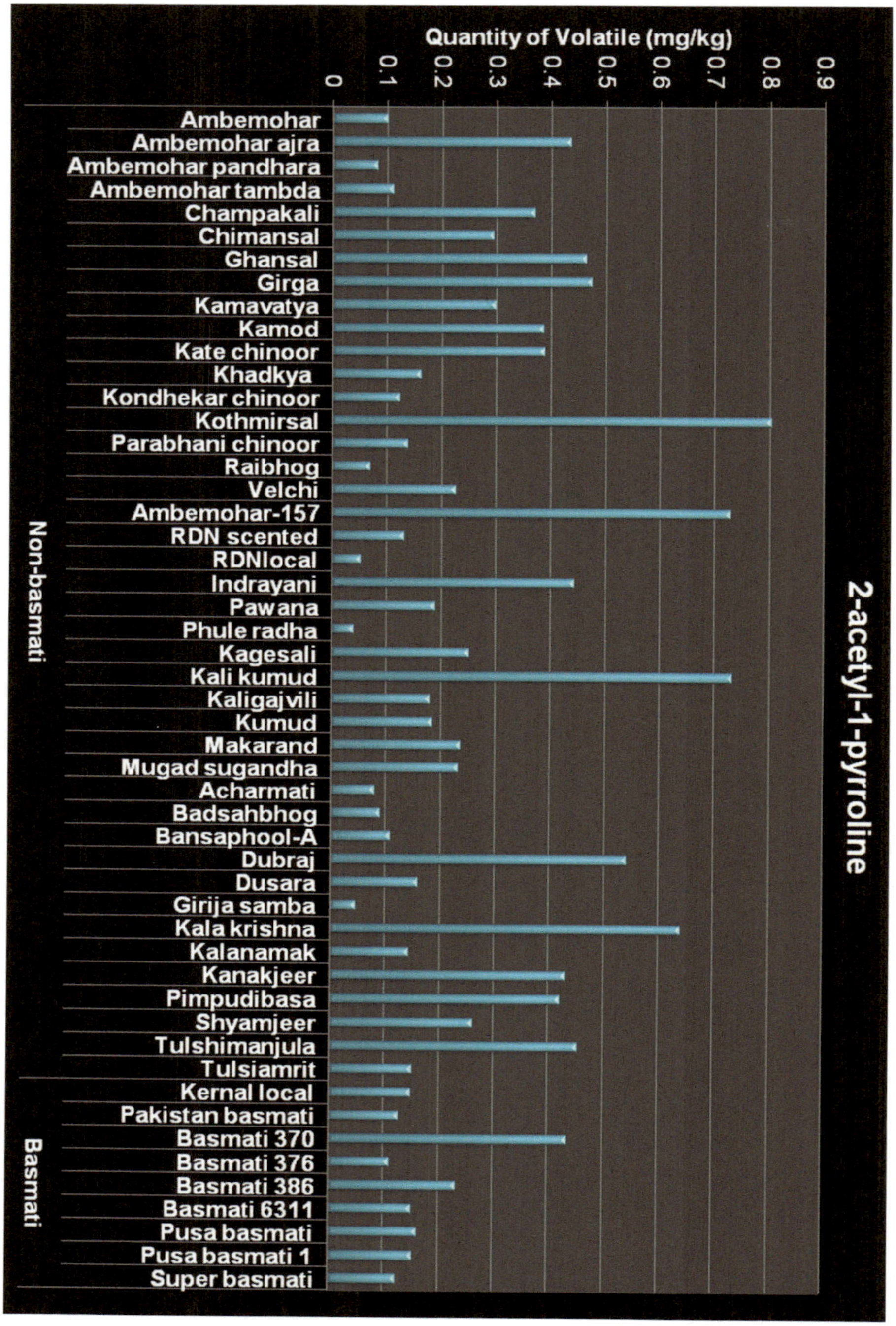

Fig. 4.12 Variation in 2-acetyl-1-pyrroline content among the cultivars under study

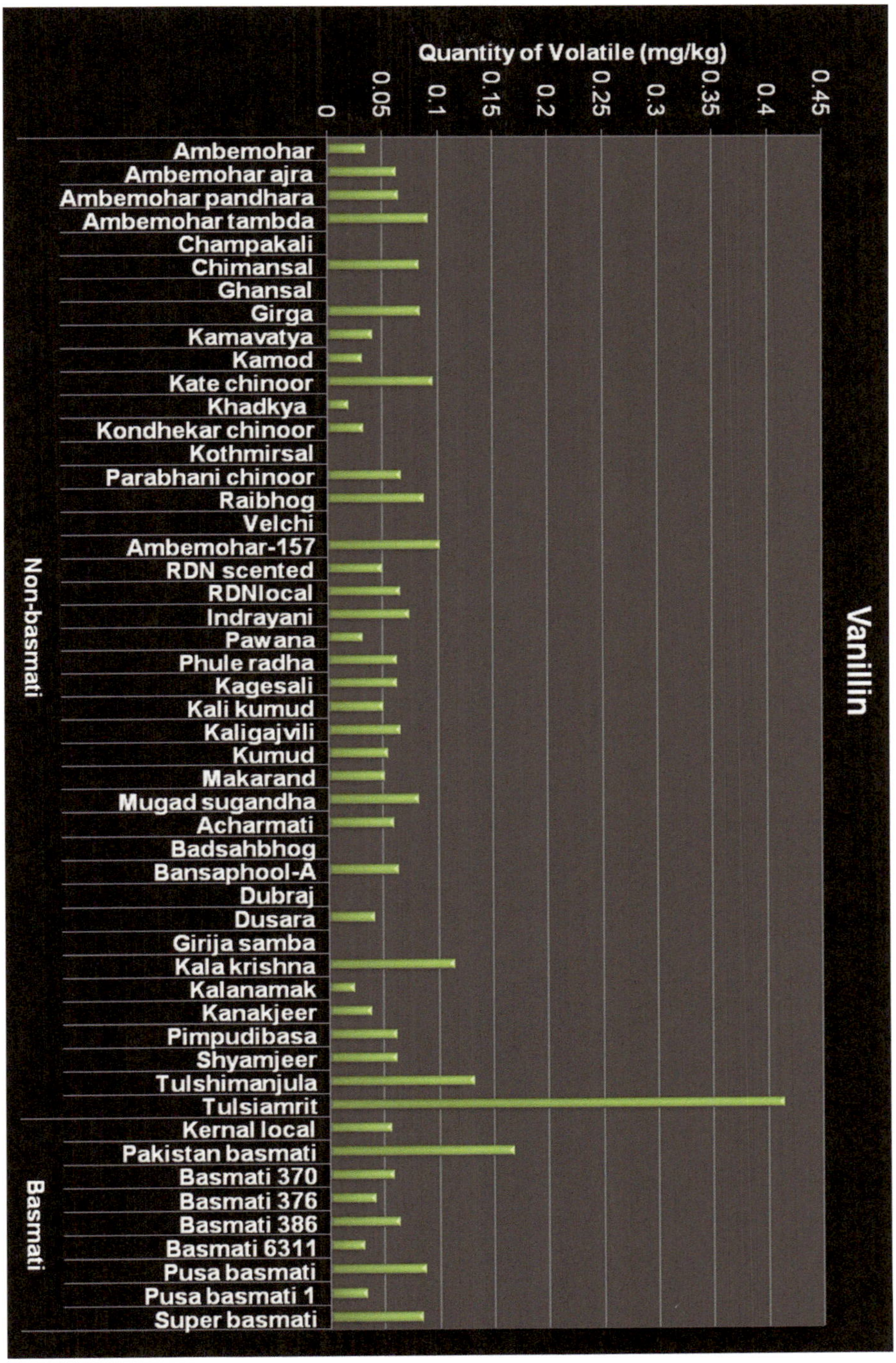

Fig. 4.13 Variation in vanillin content among the cultivars under study

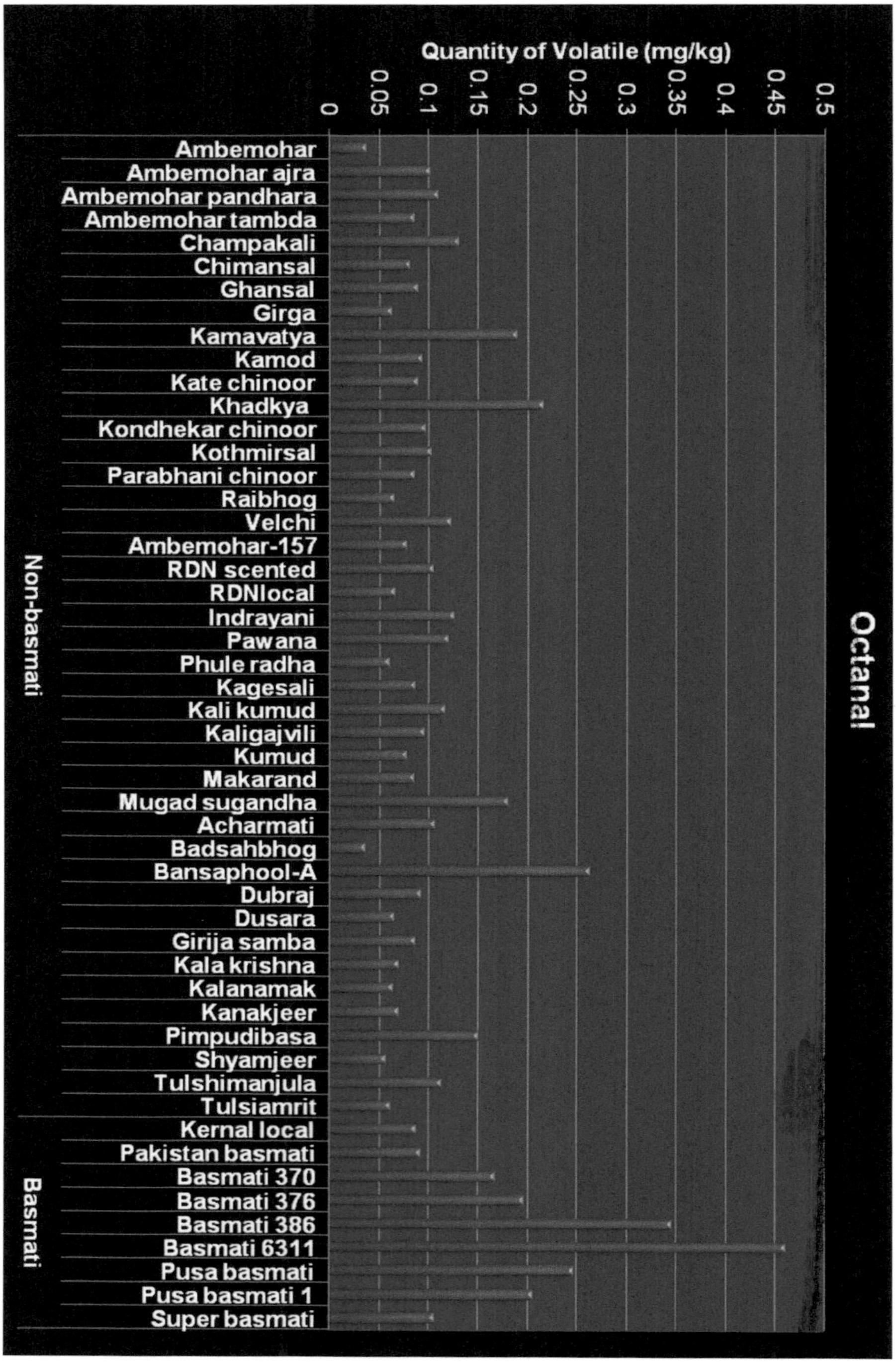

Fig. 4.14 Variation in octanal content among the cultivars under study

Fig. 4.15 Variation in hexanal content among the cultivars under study

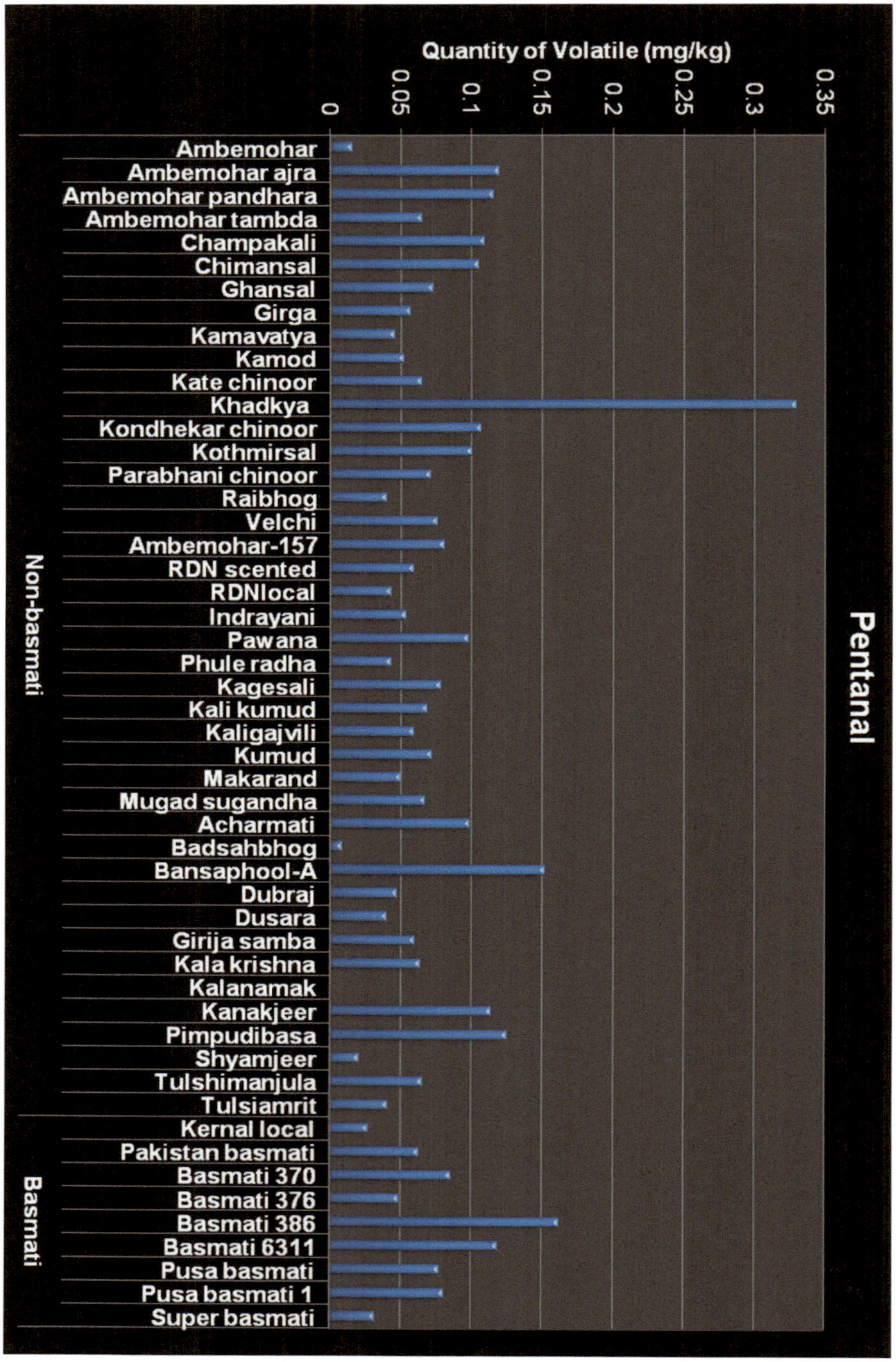

Fig. 4.16 Variation in pentanal content among the cultivars under study

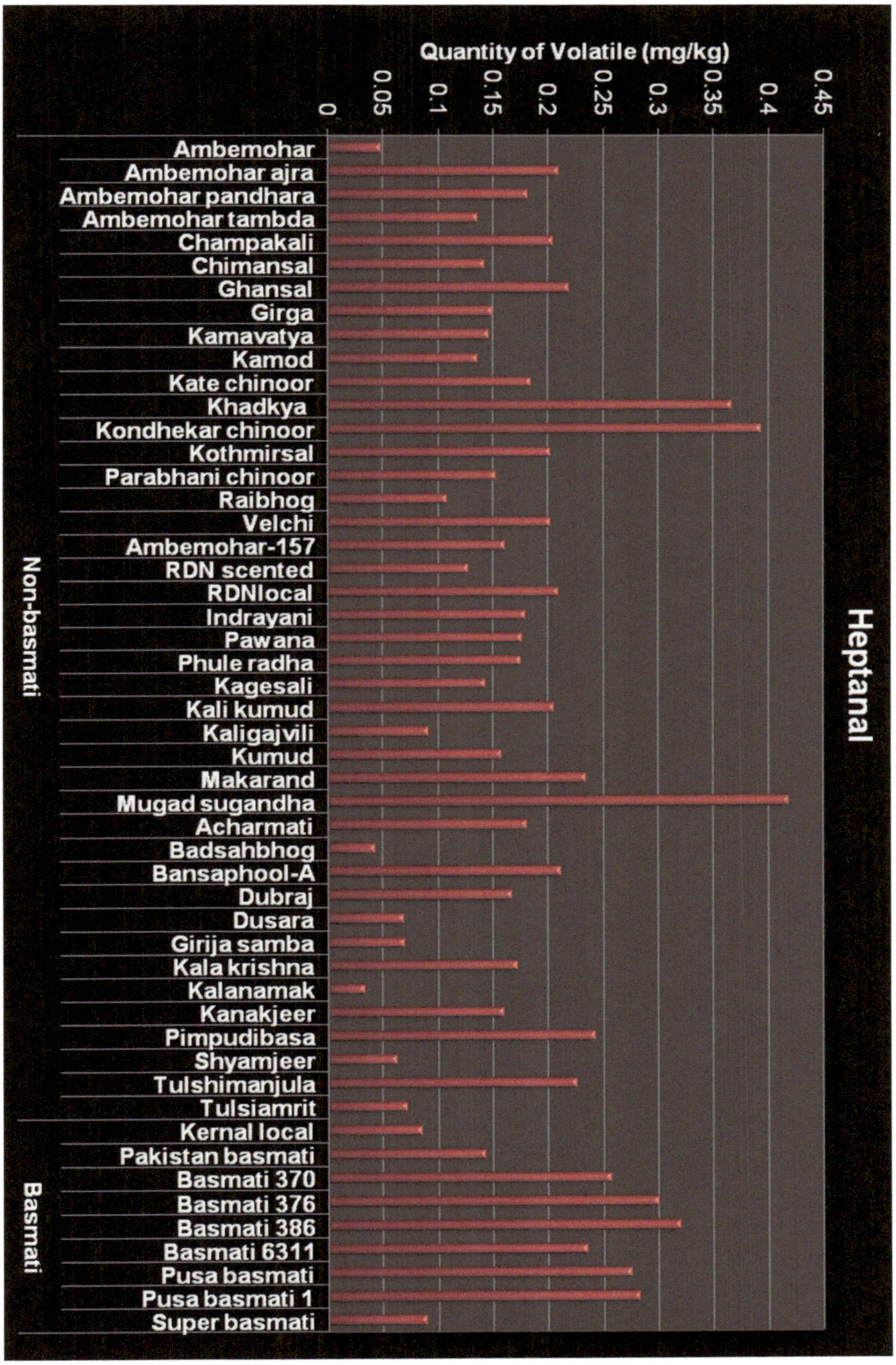

Fig. 4.17 Variation in heptanal content among the cultivars under study

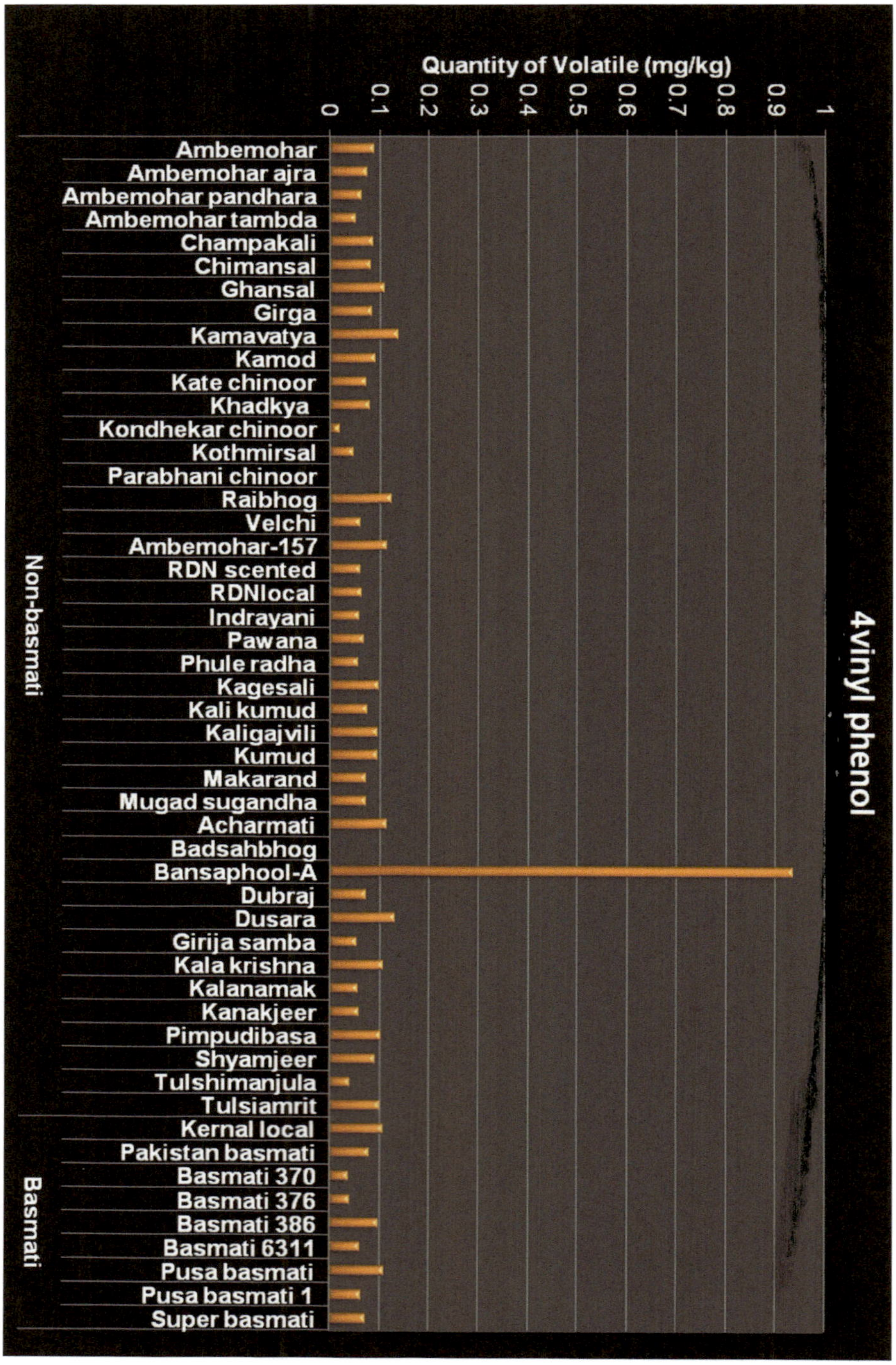

Fig. 4.18 Variation in 4 vinyl phenol content among the cultivars under study

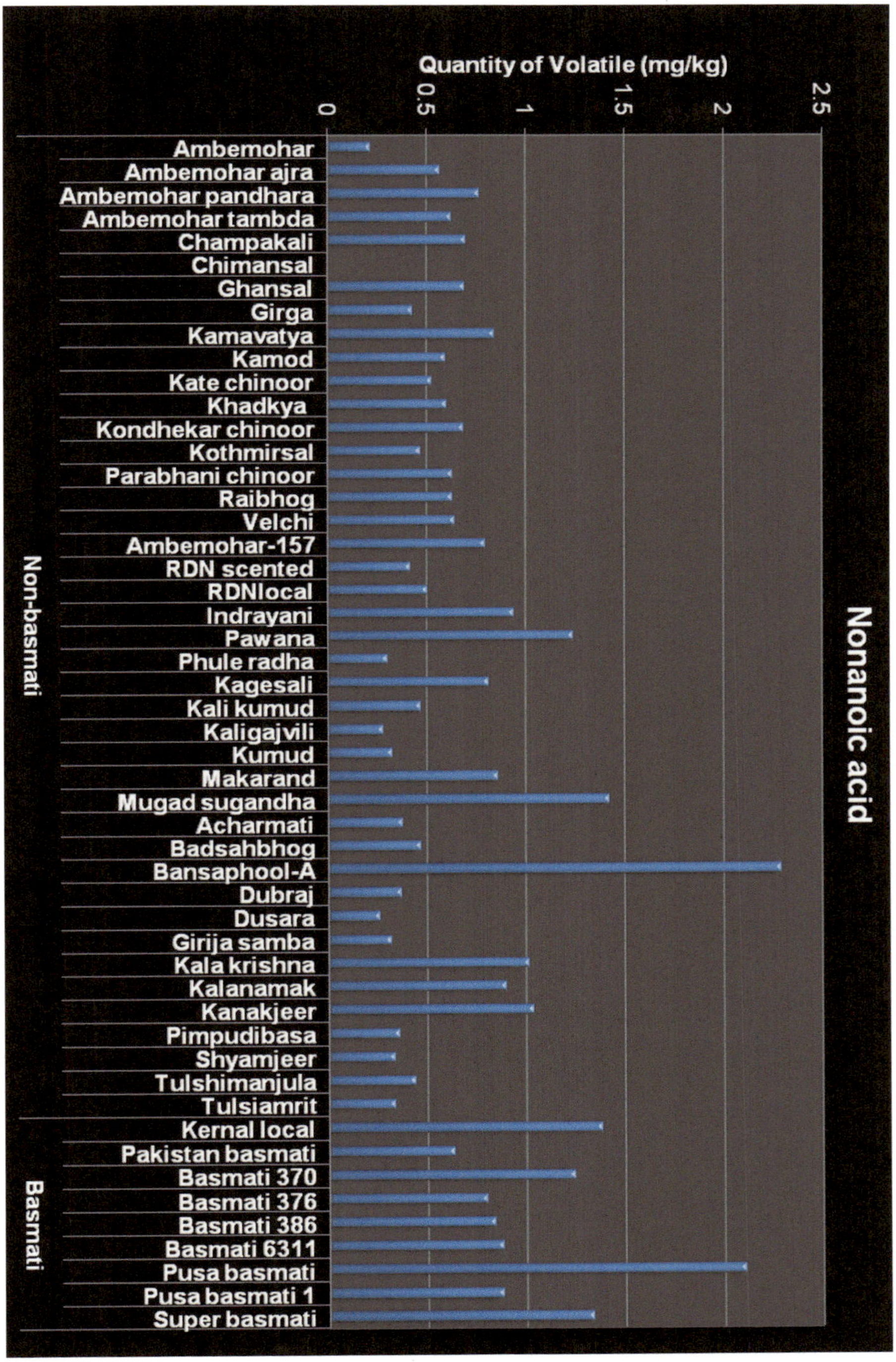

Fig. 4.19 Variation in nonanoic acid content among the cultivars under study

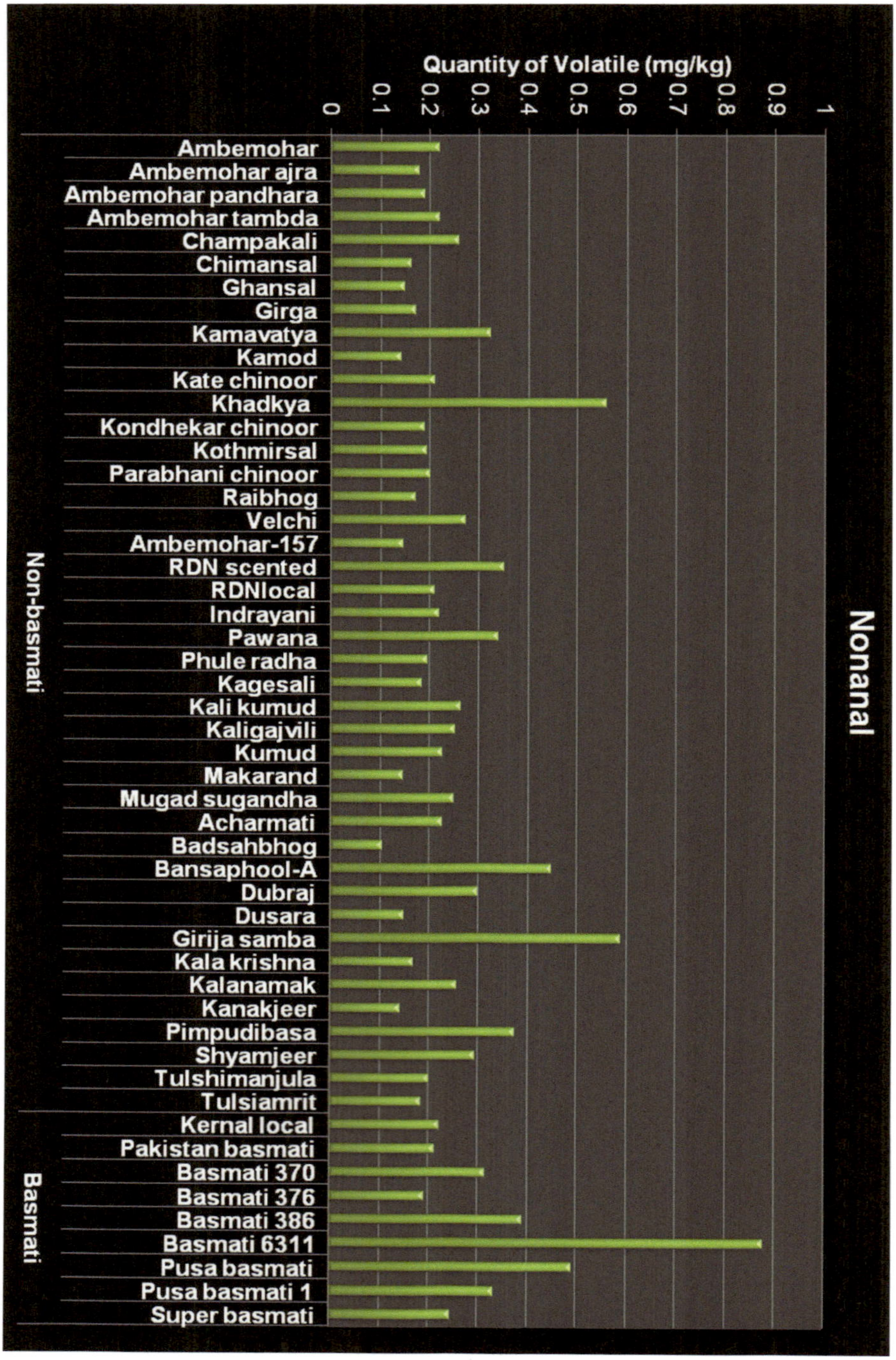

Fig. 4.20 Variation in nonanal content among the cultivars under study

Fig. 4.21 Variation in 1-hexanol content among the cultivars under study

References

Ajarayasiri J, Chaiseri S (2008) Comparative study on aroma-active compounds in Thai, black and white glutinous rice varieties. Kasetsart J (Nat Sci) 42:715–722

Arthur CL, Pawliszyn J (1990) Solid-phase microextraction with thermal desorption using fused Silica optical fibers. Anal Chem 62:2145–2148

Bergman C, Delgado J, Bryant R, Grimm C, Cadwallder K, Webb BA (2000) Rapid gas chromatographic technique for quantifying 2-Acetyl-1-Pyrroline and hexanal in rice *Oryza sativa* L. Cereal Chem 77(4):454–458

Bounphanousay C, Jaisil P, Fitzgerald M, Hamilton NRS, Sanitchon J (2008) Chemical and molecular characterization of fragrance in black glutinous rice from Lao PDR. Asian J Plant Sci 7:1–7

Bradbury LMT, Henry RJ, Jin QS, Reinke RF, Waters DLE (2005) A perfect marker for fragrance genotyping in rice. Mol Breed 16:279–283

Buttery RG, Ling LC, Juliano BO (1982) 2-acetyl-1-pyrroline: an important aroma component of cooked rice. Chem Ind 23:958

Buttery RG, Ling LC, Juliano BO, Turnbaugh JG (1983) Cooked rice aroma and 2-acetyl–1-pyroline in rice. J Agric Food Chem 31:823–826

Buttery R, Turnbaugh J, Ling L (1988) Contributions of volatiles to rice aroma. J Agric Food Chem 36:1006–1009

Buttery RG, Orts WJ, Takaoka GR, Nam Y (1999) Volatile flavor components of rice cakes. J Agric Food Chem 47(10):4353–4356

Champagne ET (2008) Rice aroma and flavor: A literature review. Cereal Chem 85(4):445–454

Frankel EN (2005) Lipid oxidation. Oily Press, Bridgewater

Grimm CC, Bergman C, Delgado JT, Bryant R (2001) Screening for 2-acetyl-1-pyrroline in the headspace of rice using SPME/GC-MS. J Agric Food Chem 49:245–249

Hamm S, Bleton J, Connan J, Tchapla A (2005) A chemical investigation by headspace SPME and GC–MS of volatile and semi-volatile terpenes in various olibanum samples. Phytochemistry 66(12):1499–1514

Jezussek M, Juliano B, Schieberle P (2002) Comparison of key aroma compounds in cooked brown rice varieties based on aroma extract dilution analysis. J Agric Food Chem 50:1101–1105

Kataoka H (2002) Automated sample preparation using in-tube solid-phase micro- extraction and its application – a review". Anal Bioanal Chem 373:31–45

Laguerre M, Mestres C, Davrieux F, Ringuet J, Boulanger R (2007) Rapid discrimination of scented rice by solid phase microextraction, mass spectrometry and multivariate analysis used as a mass sensor. J Agric Food Chem 55:1077–1083

Lam HS, Proctor A (2003) Milled rice oxidation volatiles and odor development. J Food Sci 68:2676–2681

Mahatheeranont S, Keawsa-ard S, Dumri K (2001) Quantification of the rice aroma compound, 2-acetyl-1-pyrroline, in uncooked Khao Dawk Mali 105 brown rice. J Agric Food Chem 49:773–779

Mansoor MA, Proctor A (2004) Volatile components analysis of commercially milled head and brown rice. J Food Sci 69(8):632–636

Maraval I, Mestres C, Pernin K, Ribeyre F, Boulanger R, Guichard E, Gunata Z (2008) Odor-active compounds in cooked rice cultivars from Camargue (France) analyzed by GC#O and GC#MS. J Agric Food Chem 56(13):5291–5298

Maraval I, Sen K, Agrebi A, Menut C, Morere A, Boulanger R, Gay F, Mestres C, Gunata Z (2010) Quantification of 2-acetyl-1-pyrroline in rice by stable isotope dilution assay through headspace solid-phase microextraction coupled to gas chromatography-tandem mass spectrometry. Anal Chim Acta 675(2):148–155

Mathure S, Wakte KV, Nadaf A, Jawali N (2011) Quantification of 2-acetyl-1-pyrroline and other rice aroma volatiles among Indian scented rice cultivars by HS-SPME/GC-FID. Food Anal Methods 4:326–333. doi:10.1007/s12161-010-9171-3

Mathure S, Jawali N, Thengane RJ, Nadaf AB (2014) Comparative quantitative analysis of headspace volatiles and their association with *BADH2* marker in non-basmati scented, basmati and non-scented rice (*Oryza sativa* L.) cultivars of India. Food Chem 142:383–391

Mildner-Szkudlarz S, Jelen HH, Zawirska-Wojtasiak R, Wasowicz E (2003) Application of headspace – solid phase microextraction and multivariate analysis for plant oils differentiation. Food Chem 83:515–522

Nadaf AB, Krishnan S, Wakte KV (2006) Histochemical and biochemical analysis of major aroma compound (2-acetyl-1-pyrroline) in Basmati and other scented rice (*Oryza sativa* L). Curr Sci 91(11):1533–1536

Nagaraju J, Kathirvel M, Kumar RR, Siddiq EA, Hasnain SE (2002) Genetic analysis of traditional and evolved Basmati and non-Basmati rice varieties by using fluorescence-based ISSR-PCR and SSR markers. Proc Natl Acad Sci U S A 99(9):5836–5841

Petrov M, Danzart M, Giampaoli P, Faure J, Richard H (1996) Rice aroma analysis – discrimination between a scented and a non-scented rice. Sci Aliment 16:347–360

Pico Y, Fernandez M, Ruiz MJ, Font G (2007) Current trends in solid-phase-based extraction techniques for the determination of pesticides in food and environment. J Biochem Biophys Methods 70:117–131

Sood BC, Siddiq EA (1978) A rapid technique for scent determination in rice. Indian J Gen Plant Breed 38(2):268–271

Sriseadka T, Wongpornchai S, Kitsawatpaiboon P (2006) Rapid method for quantitative analysis of the aroma impact compound, 2-acetyl-1-pyrroline, in fragrant rice using automated headspace gas chromatography. J Agric Food Chem 54:8183–8189

Stashenko EE, Martinez JR (2007) Sampling volatile compounds from natural products with headspace/solid-phase micro-extraction. J Biochem Biophys Methods 70:235–242

Steffen A, Pawliszyn J (1996) The analysis of flavor volatiles using headspace solid-phase microextraction. J Agric Food Chem 44:2187–2193

Tananuwong K, Lertsiri S (2010) Changes in volatile aroma compounds of organic fragrant rice during storage under different conditions. J Sci Food Agric 90(10):1590–1596

Tava A, Bocchi S (1999) Aroma of cooked rice (*Oryza sativa*): comparison between commercial Basmati and Italian line B5-3. Cereal Chem 76:526–529

Wakte KV, Nadaf AB, Thengane RJ, Jawali N (2010) Optimization of HS-SPME conditions for quantification of 2 acetyl-1-pyrroline in *Pandanus amaryllifolius* Roxb. Food Chem 121(2):595–600

Wercinski S (1999) Solid phase microextraction: a practical guide. CRC, Boca Raton

Widjaja RW, Craske JD, Wootton M (1996) Comparative studies on volatile components of non-fragrant and fragrant rices. J Sci Food Agric 70:151–161

Wongpornchai S, Sriseadka T, Choonvisase S (2003) Identification and quantitation of the rice aroma compound, 2-acetyl-1-pyrroline, in bread flowers (*Vallaris glabra* Ktze). J Agric Food Chem 51:457–462

Yang DK, Lee KS, Jeong OY, Kim KJ, Kays SJ (2008a) Characterization of volatile aroma compounds in cooked black rice. J Agric Food Chem 56(1):235–240

Yang DS, Robert LS, Kyu-Seong L, Stanley JK (2008b) Comparison of odor-active compounds from six distinctly different rice flavor types. J Agric Food Chem 56(8):2780–2787

Yoshihashi T (2002) Quantitative analysis on 2-Acetyl-1-pyrroline of an aromatic rice by stable isotope dilution method and model studies on its formation during cooking. J Food Sci 67(2):619–622

Zeng Z, Zhang H, Zhang T, Tamogami S, Chen JY (2009) Analysis of flavor volatiles of glutinous rice during cooking by combined gas chromatography–mass spectrometry with modified headspace solid phase microextraction method. J Food Compos Anal 22(4):347–353

Zhou Z, Robards K, Helliwell S, Blanchard C (2002) Composition and functional properties of rice. Int J Food Sci Technol 37:849–868

5 Molecular Marker-Based Assessment of Genetic Diversity and Quality Trait Specific Marker Validation

5.1 Introduction

India is considered as one of the centers of origin of rice (*Oryza sativa* L. ssp. *indica*) and a center of its diversity (Rana 1995). Archeological excavations have shown that rice has been a part of the food economy of early man in China and India from around 5000 BC (Sharma et al. 2000). Thousands of years of evolution because of adaptation to varied agroclimatic conditions and selection by farmers based on diverse quality preferences have resulted in accumulation of rich genetic diversity in rice. Singh and Singh (2003) reported that India has the world's largest germplasm of scented rice varieties. Among scented rice, basmati rice is a unique varietal group which has distinguished itself in consequence of natural and human selection. The uniqueness of basmati rice is determined by extra long slender grain, higher elongation on cooking, soft and fluffy texture of cooked rice, and pleasant aroma. Besides the genetic makeup, the qualities of basmati rice are expressed when it is grown in the northwestern foothills of Himalayas under unique combination of growing conditions such as soil, water, climate, and cultural practices (Siddiq et al. 2012).

In India, cultivation of basmati types is restricted to North India, while nonbasmati types are cultivated in several regions of India and in varied climatic conditions. Though the nonbasmati types do not possess all the characteristics of basmati, these varieties excel equally as far as aroma and other characters are concerned. Nonbasmati scented rice usually has short to medium grains. Some of the outstanding examples of short to medium grain length local scented rice are Kalanamak, Shakarchini, and Hansraj of Uttar Pradesh; Dubraj and Chinoor of Chhatisgarh; Kalajoha of the northeast; Kagisali of Karnataka; Tulsi Manjari and Katarani of Bihar; Ambemohar of Maharashtra; and Randhunipagal of Orissa and West Bengal region (Singh et al. 2000a). These varieties are popularly traded and fetch prices comparable to basmati in their native areas of cultivation where farmers and consumers are aware of their qualities. Hence, there is a keen interest to identify the genetic variation among basmati and nonbasmati scented types with specific interest in landraces.

5.1.1 Molecular Markers for Diversity Assessment in Rice

DNA markers based on variation in the genome sequence help in the detection of genetic variation among different individuals. In the past few decades, several DNA-based markers are developed. Advances in polymerase chain reaction (PCR) technology has provided a sizable number of useful molecular markers over hybridization-based restriction fragment length polymorphism

A. Nadaf et al., *Scented rice (Oryza sativa L.) Cultivars of India: A Perspective on Quality and Diversity*, DOI 10.1007/978-81-322-2665-9_5

(RFLP) for genome analysis and mapping (Kumar 1999; Semagn et al. 2006; Mondini et al. 2009). In 1995, Vos et al. described AFLP as a robust and reliable technique for DNA fingerprinting as a means of precise genotyping. AFLP can be generated from DNA without prior knowledge of the genome sequence. AFLP combines the power of RFLP with the flexibility of PCR-based technology (Vos et al. 1995). The number of amplicons per AFLP assay is a function of the number selective nucleotides in the AFLP primer combination, the selective nucleotide motif, GC content, and physical genome size and complexity (Mondini et al. 2009).

In rice, both random amplified polymorphic DNA (RAPD) and AFLP markers are found to be effective in the assessment of genetic diversity (Fuentes et al. 1999) and gave similar segregation of cultivars in subspecies by UPGMA analysis (Mackill et al. 1996). McGregor et al. (2000) compared the reproducibility factor of different marker systems and reported simple sequence repeats (SSRs) (100 %) to be highly reproducible, followed by AFLP (99.6 %), intersimple sequence repeats (ISSRs), (87 %) and RAPD (84.4 %), while Saini et al. (2004) recorded that AFLP has higher discriminatory power over SSR and ISSR markers and can be used in distinguishing basmati and nonbasmati rice. Fuentes et al. (1999) reported higher efficiency of AFLP to detect polymorphism over RAPD and isozymes in Cuban rice. AFLP markers are effective in the detection of polymorphism between closely related species and to infer phylogenetic relationships (Mueller and Wolfenbarger 1999). Ovesna et al. (2002) reported that the AFLP technology is reported as a powerful tool for screening of biodiversity, fingerprinting studies, and evaluation of genetic variation in germplasm collections. AFLP is also reported as a genetic tool for identification of genetic lineages (Aggarwal et al. 2002), identification of basmati rice mutants for desirable agronomic and quality characters (Rashid et al. 2009), and varietal identification among Indian rice cultivars (Singh et al. 1999). Therefore, AFLP was preferred in the present study for diversity analysis in scented rice.

Singh et al. (1999) attempted high resolution of DNA fingerprinting by AFLP and segregated 42 Indian rice varieties in four clusters. Their study revealed higher polymorphism and detected unique loci in landraces. In 2000, Joshi et al. studied genetic diversity and phylogenetic relationship in genus *Oryza* by ISSR. Fluorescent-labeled rice microsatellite markers to survey allelic diversity were tested on 52 basmati and nonbasmati rice collected from parts of India that indicated diversity in basmati and nonbasmati accessions (Jain et al. 2004). Eighteen accessions of Hansraj (basmati type rice) generated 95.1 % polymorphism by RAPD (Raghunathachari et al. 2000). Davierwala et al. (2000) estimated genetic diversity of Indian elite rice varieties with three different markers. Raychaudhari et al. (2001)) utilized RAPD-based DNA fingerprinting for the identification and classification of scented rice genotypes, using 58 primers which clustered short grain scented rice separately from basmati. Genetic diversity among 49 Indian accessions of rice (*Oryza sativa* ssp. *indica*), including 29 landraces from Jeypore, 12 modern cultivars, and eight traditional cultivars from Tamilnadu, was investigated by Prashanth et al. (2002) using AFLP markers. They reported that the process of breeding modern cultivars did not cause significant genetic erosion in rice landraces under cultivation. Nagaraju et al. (2002) performed a study based on ISSR-PCR and SSR markers to assess the genetic relationship between traditional and evolved basmati and nonbasmati rice varieties. Singh et al. (2006) recorded that on the basis of RAPD, Kalanamak accessions were diverse from evolved and traditional basmati varieties. Joshi and Behera (2006) used microsatellite markers for the identification and differentiation of indigenous nonbasmati scented rice genotypes of India. Bhuyan et al. (2007) reported that indigenous rice cultivars of Assam were distinctly clustered as scented *joha* and nonscented cultivars by RAPD and ISSR markers. Sivaranjani et al. (2010) assessed genetic diversity among basmati and nonbasmati aromatic landraces of India using SSR to classify them into two major clusters.

Maharashtra state has a good number of aromatic landraces under cultivation. In spite of this, the diversity of these cultivars has remained unexplored. Hence, to address the need for diversity, analysis of rice cultivars grown in Maharashtra and Karnataka and their comparison with other nonbasmati scented rice, nonscented rice, and basmati rice cultivars from India have been carried out.

5.1.2 Validation of Molecular Markers Linked with Quality Traits in Scented Rice

New developments in genomics have provided tools for discovering and tagging novel alleles and offer great promise to understand the inheritance of grain quality traits and crop improvement through their use in marker-assisted selection (MAS) (Xu and Crouch 2008). The selection of target traits can be achieved by using molecular markers developed from the actual gene sequences or markers that are closely linked to genes or to major quantitative trait locus (QTL). Earlier it was assumed that most markers associated with QTLs from preliminary mapping studies were directly useful in MAS. However, in recent years, it has become widely accepted that QTL confirmation and QTL validation may be required (Langridge et al. 2001). Marker validation involves testing the reliability of marker in determining the target phenotype in independent populations and different genetic backgrounds (Collard et al. 2005).

In rice, several molecular markers have been developed and found as suitable in MAS for resistance to or tolerance to blast, bacterial blight, virus diseases, brown plant hopper, drought, submergence, salinity, and low temperature and improved agronomic and grain quality traits (Jena and Mackill 2008). Grain quality plays a vital role in rice variety development and is considered a major breeding objective along with yield (Juliano and Duff 1991). Grain quality includes multifaceted parameters such as grain appearance, milling quality, cooking and eating quality, nutritional quality, and aroma in case of specialty rice. The present study addresses the need for validation of markers available for aroma, amylose content, and grain length in Indian rice cultivars.

5.1.3 Molecular Markers for Aroma

In Asia, aroma is considered as one of the important rice quality traits. 2-acetyl-1-pyrroline (2AP) is the key volatile compound which imparts the characteristic popcorn-like fragrance to scented rice. In 1992, Ahn et al. made systematic efforts to map fragrance trait in Della (Jasmine-derived scented variety) rice and concluded that the aroma gene is linked to the RFLP clone RG28 on chromosome 8, at a distance of 4.5 cM. A major quantitative trait locus (QTL) exhibiting correlation with 2AP content was tagged by Lorieux et al. (1996) in the same region. This locus was further confirmed by Wanchana et al. (2005), Bradbury et al. (2005a), Chen et al. (2006)), and Vanavichit (2007). In 2005, Bradbury et al. (2005a) identified eight base pair (bp) deletion and three SNPs in exon 7 of the gene encoding betaine aldehyde dehydrogenase 2 (*badh2*) on chromosome 8; as a consequence, a premature stop codon results in loss of betaine aldehyde dehydrogenase 2 activity as the likely cause of fragrance in Jasmine and Basmati type rice. Several linked microsatellite markers (Garland et al. 2000; Cordeiro et al. 2002) and single nucleotide polymorphism (Jin et al. 2003) are available to discriminate fragrant and nonfragrant varieties. As these markers are linked and do not allow to determine actual fragrance status, Bradbury et al. (2005b) developed a single tube allele specific amplification (ASA) assay, which allowed differentiation between fragrant and nonfragrant and between homozygous and heterozygous. Amarawathi et al. (2008) validated ASA assay and recommended primers nksbad2F and nksbad2R designed from the conserved sequences flanking the 5+3 bp deletion in the seventh exon as reproducible markers for aroma. These primers consistently amplified 82 bp fragment for scented and 90 bp fragment for nonscented varieties (Amarawathi et al. 2008). A study on composition of headspace volatiles and validation of marker available for aroma locus among indigenous scented cultivars of India was carried out by our group help in understanding locus-specific diversity among these rice cultivars (Mathure et al. 2014). Kovach et al (2009) established a single origin of the *badh2.1* allele within the *japonica* varietal group and demonstrated the introgression of this allele from *japonica* to *indica*. However, Shao et al (2013) stated that original *badh2* mutations happened in the *indica* and *japonica* types in the process of rice domestication, and then *badh2* genes have transferred to each other with the help of the artificial selection. Occurrence of genetic mutations across the divergent subpopulations of *O. sativa* remains as a central question in rice evolutionary biology.

5.1.4 Molecular Markers for Amylose Content

Amylose content is a key determinant of the cooking and processing qualities of rice (Juliano 1985). Waxy rice which lacks in amylose is sticky and remains firm when cooked, while rice with high amylose cooks dry and hard. Rice with intermediate amylose content cooks moist and tender and does not become hard on cooking. Majority of the scented rice possesses intermediate amylose (Sood et al. 1983) and is preferred for cooking qualities, along with aroma. Among several enzymes involved in the starch synthesis, viz., ADP-glucose pyrophosphorylase, isoforms of starch-branching enzyme, starch synthase, and debranching enzyme, synthesis of amylose is catalyzed by the granule-bound starch synthase (*GBSS*) encoded by the *Wx* gene (also called as *waxy* gene) (Shure et al. 1983; Smith et al. 1997). Several researchers have assigned *Wx* region located on chromosome 6 in rice as a major QTL which primarily controls amylose content (He et al. 1999; Tan et al. 1999; Lanceras et al. 2000; Septiningsih et al. 2003; Tian et al. 2005; Fan et al. 2005; Wang et al. 2007). In 1990, presence of 13 introns and 14 exons in *Wx* gene, along with the entire gene sequence, was reported by Wang et al. (1990). Wang et al. (1995) further reported that that the amylose content in rice endosperm is related to the posttranscriptional regulation of the *waxy* gene. They recorded that rice cultivars with high amylose content produce large amounts of mature *Wx* mRNA, glutinous rice produces a large *Wx* transcript containing intron 1 instead of mature *Wx* mRNA, while cultivars with intermediate levels of amylose produce both the mature *Wx* mRNA and the incompletely spliced *Wx* pre-mRNA with intron 1 still present. Wang et al. (1995) also suggested a role of substitution at the putative leader intron 5′-splice site to alter the efficiency of excision of intron 1 from the *Wx* transcript. Bligh et al. (1995) identified a polymorphic microsatellite in the form of $(CT)_n$ repeats located 55 bp upstream of the putative 5′-leader intron splice site as marker for starch quality closely linked to the *waxy* gene. In a study by Ayres et al. (1997) on nonglutinous rice cultivars of US rice germplasm, *Wx* microsatellite classes explained 82.9 % variation in apparent amylose content. This was further confirmed by Shu et al. (1999) and Bergman et al. (2001) in nonwaxy rice. Few reports on the presence of SNPs in *Wx* locus and resultant variation in the apparent amylose content are also available (Ayres et al. 1997; Larkin and Park 2003). Tan and Zhang (2001) studied SSRs present in the leader region of the *waxy* gene and reported a perfect correlation between repeat number of SSR and amylose content. A systematic effort was made by Bao et al. (2006) to detect polymorphism in starch-synthesizing genes and to clarify ability of *Wx* microsatellites (*Wx*-SSR) to differentiate rice on the basis of physicochemical quality groups. They recorded that *Wx*-SSR is useful in marker-assisted breeding for the improvement of starch quality of rice as it explained nearly 90 % of the total variation in apparent amylose content in nonwaxy rice obtained from China. He et al. (2006) employed cleaved amplified polymorphic sequences (CAPS) marker for *Wx* gene and reported that the marker accounted for 58.5 % of the variation in apparent amylose content. The $(CT)_n$ microsatellite has been widely studied in cultivars and breeding lines from the US (Ayres et al. 1997; Bergman et al. 2001), Thailand (Prathepha and Baimai 2004) and China (Bao et al. 2002, 2006), and Iran (Rahemi et al. 2007). However, no study has been undertaken on Indian cultivars except for a recent study on Njavara, Jyothi, and IR 64 by Deepa et al. (2009).

5.1.5 Molecular Markers for Grain Length

Appearance of milled rice is one of the vital traits influencing consumer preference and is considered as one of the criteria in breeding of quality rice (Adair et al. 1966). Grain appearance depends on the size and shape of kernel and translucency and chalkiness of grain. Size of the grain is determined by grain length and is quantitatively inherited (Mckenzie and Rutger 1983). Lin (1995) reported QTLs grain length on chromosome 7 in *indica* rice by RFLP mapping. Huang et al. (1997) identified four QTLs for grain length located on chromosomes 1, 3, and 10 of which QTL on chromosome 1 explained maximum variation in grain length from a cross between an *indica* variety IR64 and a *japonica* variety Azucena. Redona and Mackill (1998) identified seven QTLs on chromosomes 2, 3, 4, 7, and 10 in a tropical *japonica* × *indica* cross. In interspecific cross of *O. sativa* × *O. glaberrima*, QTLs for grain length were detected on chromosomes 3 and 6 (Aluko et al. 2004). Ge et al. (2005) detected QTLs for grain length on chromosomes 2 and 3. Wan et al. (2006) evaluated six main-effect QTLs in eight environments and detected a stable QTL on chromosome 3, along with other QTLs on chromosomes 2, 5, 7, and 9. Wang et al. (2006) reported that eight of the 12 rice chromosomes harbored QTLs affecting grain length, among which three major stable QTLs located on chromosomes 1, 3, and 4 were detected in all eight environments using chromosome segment substitution (CSS) lines. Grain length is mostly controlled by the major QTLs on chromosome 3 (Kubo et al. 2001; Xing et al. 2001; Li et al. 2004, Wan et al. 2005, 2006). Fan et al. (2006) reported *GS3* as a major QTL for grain length located on chromosome 3 which encodes a putative transmembrane protein. Though QTL for grain length is assessed in several genetic backgrounds, no study was available on basmati cultivars which are premium scented rice cultivars of India. Amarawathi et al. (2008) assessed mapping population involving Pusa 1121, an evolved basmati cultivar, and reported a major QTL for grain length on chromosome 1, along with two QTLs on chromosome 7. Major QTL for grain length is located on chromosome 1 in populations with *indica* background by Huang et al. (1997), Wang et al. (2006), and (Amarawathi et al. 2008). RM 431 is tightly linked with QTL for grain length on chromosome 1 with a map distance of 0.1 cM which can be used in MAS after validation (Amarawathi et al. 2008).

Hence, this study was undertaken to assess the utility of *badh2* marker (Amarawathi et al. 2008), *Wx*-SSR (Bligh et al. 1995), and RM 431 in marker-assisted selection for aroma, apparent amylose content, and grain length, respectively, among diverse Indian rice cultivars with specific interest in nonbasmati scented rice cultivars.

5.2 Methodology and Approaches for Genetic Diversity Assessment

5.2.1 Plant Material

Fifty-three rice (*Oryza sativa* L. spp *indica*) cultivars representing nonbasmati scented, basmati, and nonscented rice cultivars were selected for the study (Table 5.1). These rice cultivars were available at the Dept. of Botany, University of Pune, Pune, Maharashtra, India either as a personal collection from Maharashtra and Karnataka or as a germplasm from rice research stations and institutes, viz., Agricultural Rice Research Station, Radhanagari (Kolhapur, Maharashtra, India); Rice Research Station, Shindevahi (Chandrapur, Maharashtra, India); Rice Research Station, Karjat (Raigadh, Maharashtra, India); Agricultural Research Station, Sirsi (Karwar, Sirsi, Karnataka, India); Rice Research Station, Mugad (Dharwad, Karnataka, India); Indian Agricultural Research Institute (New Delhi, India); and National Seed Corporation Ltd (New

Table 5.1 Details of rice (*Oryza sativa* L.) cultivars used for study

Sr. No.	Cultivar	Abbreviation	Type	Locality/Source
1	Ambemohar	Amb	NB, Sc, L	Pune, Maharashtra
2	Ambemohar Ajra	AmbAjra	NB, Sc, L	Kolhapur, Maharashtra
3	Ambemohar Pandhara	AmbPandhara	NB, Sc, L	Pune, Maharashtra
4	Ambemohar-Tambda	AmbTambda	NB, Sc, L	Pune, Maharashtra
5	Champakali	Champakali	NB, Sc, L	Kolhapur, Maharashtra
6	Chimansal	Chimansal	NB, Sc, L	Nashik, Maharashtra
7	Gham	Gham	NB, Sc, L	Raigadh, Maharashtra
8	Ghansal	Ghansal	NB, Sc, L	Kolhapur, Maharashtra
9	Girga	Girga	NB, Sc, L	Kolhapur, Maharashtra
10	Jiri	Jiri	NB, Sc, L	Pune, Maharashtra
11	Kala bhat	Kalabhat	NB, Sc, L	Ahmadnagar, Maharashtra
12	Kalsal	Kalsal	NB, Sc, L	Pune, Maharashtra
13	Kamavatya	Kamavatya	NB, Sc, L	Pune, Maharashtra
14	Kamod	Kamod	NB, Sc, L	Nashik, Maharashtra
15	Kate chinoor	KateChin	NB, Sc, L	Chandrapur, Maharashtra
16	Khadkya	Khadkya	NB, Sc, L	Pune, Maharashtra
17	Kondhekar chinoor	KondhekarChin	NB, Sc, L	Chandrapur, Maharashtra
18	Kothmirsal	Kothmirsal	NB, Sc, L	Kolhapur, Maharashtra
19	Lal dodki	Laldodki	NB, Sc, L	Pune, Maharashtra
20	Raibhog	Raibhog	NB, Sc, L	Pune, Maharashtra
21	Velchi	Velchi	NB, Sc, L	Ratnagiri, Maharashtra
22	Ambemohar-157	Amb-157	NB, Sc, S	Pune, Maharashtra
23	Indrayani	Indrayani	NB, Sc, L	Pune, Maharashtra
24	Basumati	Basumati	NB, Sc, L	Belgaum, Karnataka
25	Gandhesale	Gandhesale	NB, Sc, L	Agri. Research Station, Sirsi, Karnataka
26	Geerige sanna	Geerigesanna	NB, Sc, L	Agri. Research Station, Sirsi, Karnataka
27	Kagisali	Kagisali	NB, Sc, L	Rice Research Station, Mugad, Karnataka
28	Kaligajvili	Kaligajvili	NB, Sc, L	Rice Research Station, Mugad, Karnataka
29	Kali kumud	Kalikumud	NB, Sc, L	Belgaum, Karnataka
30	Kumud	Kumud	NB, Sc, L	Belgaum, Karnataka
31	Medhini sanna bhatta	Medhinisannabt	NB, Sc, L	Agri. Research Station, Sirsi, Karnataka
32	Mysore sanna	Mysoresanna	NB, Sc, L	Agri. Research Station, Sirsi, Karnataka
33	Sanna bili bhatta	Sannabilibt	NB, Sc, L	Agri. Research Station, Sirsi, Karnataka
34	Vasane sanna bhatta	Vasanesannabt	NB, Sc, L	Agri. Research Station, Sirsi, Karnataka
35	Acharmati	Acharmati	NB, Sc, L	Rice Research Station, Shindevahi, Maharashtra
36	Badsahbhog	Badshahbhog	NB, Sc, L	Rice Research Station, Mugad, Karnataka

(continued)

Table 5.1 (continued)

Sr. No.	Cultivar	Abbreviation	Type	Locality/Source
37	Bansaphool-A	Bansphool-A	NB, Sc, L	Agri. Rice Research Station, Radhanagari, Maharashtra
38	Dubraj	Dubraj	NB, Sc, L	Rice Research Station, Shindevahi, Maharashtra
39	Dusara	Dusara	NB, Sc, L	Rice Research Station, Shindevahi, Maharashtra
40	Girija samba	Girijasambha	NB, Sc, L	Rice Research Station, Shindevahi, Maharashtra
41	Kalanamak	Kalanamak	NB, Sc, L	Rice Research Station, Shindevahi, Maharashtra
42	Shyamjeer	Shyamjeer	NB, Sc, L	Agri. Rice Research Station, Radhanagari, Maharashtra
43	Kalakrishna	Kalakrishna	NB, Sc, L	Agri. Rice Research Station, Radhanagari, Maharashtra
44	Kothmbiri	Kothmbiri	NB, Sc, L	Agri. Rice Research Station, Radhanagari, Maharashtra
45	Barke bhat	Barkebhat	NB, Sc, L	Agri. Rice Research Station, Radhanagari, Maharashtra
46	Kolamb	Kolamb	NB, NSc, L	Pune, Maharashtra
47	Manila	Manila	NB, NSc, L	Belgaum, Karnataka
48	Jaya	Jaya	NB, NSc, H	Rice Research Station, Karjat, Maharashtra
49	Kernal local	Kernallocal	Bas, Sc, L	Rice Research Station, Karjat, Maharashtra
50	Pakistan basmati	Pakistanbasmati	Bas, Sc, L	Rice Research Station, Shindevahi, Maharashtra
51	Basmati 370	Basmati370	Bas, Sc, S	Indian Agricultural Research Institute, New Delhi
52	Basmati 6311	Basmati6311	Bas, Sc, S	Rice Research Station, Karjat, Maharashtra
53	Pusa basmati-1	Pusabasmati-1	Bas, Sc, H	National Seed Corporation Ltd, New Delhi

NB – nonbasmati, Bas – basmati, Sc – scented, NSc – nonscented, L – landrace, S –selection
H-Hybrid

Delhi, India). Genomic DNA was isolated from leaves of two-week old rice seedlings (Mathure et al. 2014).

5.2.2 AFLP Analysis

The AFLP analysis was performed following the method of Vos et al. (1995). The *EcoRI* and *MseI* adapters used are given in Table 5.2.

5.2.3 Data Analyses

DNA bands from the AFLP profiles were scored as '1' for presence and '0' for absence. The light bands were excluded, and only the scorable, intense bands were considered for further analysis. Distance matrix, genetic relationship among the cultivars, and boot strap values were calculated using the TREECON software (version 1.3b) (Van de Peer and De Wachter (1993) by

Table 5.2 Details of adapters and primers used in AFLP analysis

Adapter/Primer	Sequence	Length (bp)
Double-stranded adapter		
EcoRI-a	5′-CTCGTAGACTGCGTACC-3′	17
EcoRI-b	5′-AATTGGTACGCAGTCTAC-3′	18
MseI-a	5′-GACGATGAGTCCTGAG-3′	16
MseI-b	5′- TACTCAGGACTCAT-3′	14
Preselective amplification primers		
E_A	5′-GACTGCGTACCAATTCA-3′	17
M_C	5′-GATGAGTCCTGAGTAAC-3′	17
Selective amplification primers		
E_{ACT}	5′- GACTGCGTACCAATTCACT-3′	19
E_{AAC}	5′- GACTGCGTACCAATTCAAC-3′	19
E_{ACC}	5′- GACTGCGTACCAATTCACC-3′	19
E_{AAG}	5′- GACTGCGTACCAATTCAAG-3′	19
E_{ACA}	5′- GACTGCGTACCAATTCACA-3′	19
E_{AGG}	5′- GACTGCGTACCAATTCAGG-3′	19
E_{ACG}	5′- GACTGCGTACCAATTCACG-3′	19
E_{AGT}	5′- GACTGCGTACCAATTCAGT-3′	19
E_{AGC}	5′- GACTGCGTACCAATTCAGC-3′	19
M_{CAC}	5′- GATGAGTCCTGAGTAACAC-3′	19
M_{CAA}	5′- GATGAGTCCTGAGTAACAA-3′	19
M_{CTG}	5′- GATGAGTCCTGAGTAACTG-3′	19
M_{CAG}	5′- GATGAGTCCTGAGTAACAG-3′	19
M_{CAT}	5′- GATGAGTCCTGAGTAACAT-3′	19
M_{CTA}	5′- GATGAGTCCTGAGTAACTA-3′	19
M_{CTC}	5′- GATGAGTCCTGAGTAACTC-3′	19
M_{CTT}	5′- GATGAGTCCTGAGTAACTT-3′	19
M_{CAA}	5′- GATGAGTCCTGAGTAACAA-3′	19

unweighted pair group method with arithmetic mean (UPGMA). Genetic distances between the cultivars were estimated following Nei and Li method (Nei and Li 1979). The relationship among 53 rice cultivars was portrayed graphically in the form of a dendrogram.

The polymorphism information content (PIC) per band was calculated using the following formula:

$$PIC = 1 - \Sigma P_{ij}^{2}$$

where 'P_{ij}' is the frequency of 'jth' allele for marker i and the summation extends over 'n' alleles. Average PIC per primer was deduced by averaging PIC values of all the bands generated by respective primer.

Molecular marker validation

5.2.4 Primers Used for Validation of Markers

Primers used for validation are presented in Table 5.3.

5.2.5 Data Analyses

Grain length, aroma, and amylose content determined in physicochemical analysis and aroma analysis were used for validation of markers. Alleles were scored as '1' for presence and '0' for absence of band. Pearson's correlation coefficients were calculated between band size of aroma allele and quantity of 2AP. Duncan's multiple range test was performed on average values for apparent amylose content for each allele. The

Table 5.3 Details of primers used for validation of markers

Primer	Sequence	Length (bp)	References
nksbad2 F	GGTTGCATTTACTGGGAGTTATG	23	Amarawathi et al. (2008)
nksbad2 R	TCCACAGAAATTTGGAAACAAAC	23	
Wx-SSR 484	CTTTGTCTATCTCAAGACAC	20	Bligh et al. (1995), Ayres et al. (1997), Larkin et al. (2003), Liu et al. (2004), Bao et al. (2006)
Wx-SSR 485	TTGCAGATGTTCTTCCTGATG	21	
RM 431 F	TCCTGCGAACTGAAGAGTTG	20	Temnykh et al. (2001), Amarawathi et al. (2008)
RM 431 R	AGAGCAAAACCCTGGTTCAC	20	

polymorphism information content (PIC) per allele was calculated as described above. Pearson's correlation coefficients (r) between apparent amylose content and *Wx*-SSR alleles and grain length and RM 431 alleles were estimated. All the analysis was performed using SPSS software (version 11.5, Chicago, USA).

5.3 AFLP Analysis

The AFLP analysis of 53 cultivars was carried out using ten primer combinations (Table 5.2).

The ten primer combinations produced 331 scorable bands, including 240 polymorphic bands, with an average polymorphism of 72.51 %. The highest percentage of polymorphism was obtained with $E_{ACA}+M_{CTA}$ (86.21 %) and then with $E_{ACG}+M_{CTT}$ (85.19 %). Primer combinations $E_{ACT}+M_{CAG}$ and $E_{AAG}+M_{CAT}$ generated the highest number of polymorphic bands (30 each), while $E_{AAC}+M_{CAA}$ produced the lowest number (16) of bands. Average PIC per primer combination varied in a narrow range from 0.28 ($E_{AAC}+M_{CAA}$ and $E_{AGG}+M_{CTC}$) to 0.35 ($E_{AAG}+M_{CAT}$). A representative AFLP profile obtained using primer combination $E_{ACT}+M_{CAG}$ is shown in Fig. 5.1.

As seen from Table 5.4, the genetic dissimilarity among 53 cultivars ranged from 11.28 to 54.21 %. Among nonbasmati scented rice cultivars, genetic dissimilarity varied from 11.28 % (Ambemohar and Ambemohar pandhara) to 54.21 % (Champakali and Kaligajvili). Basmati 370 and Basmati 6311 recorded genetic dissimilarity of 14.45 %, while Kernal local and Pusa basmati-1 recorded higher dissimilarity (30.36 %) among basmati types. Genetic dissimilarity among nonscented rice cultivars ranged from 20.35 % (Manila and Jaya) to 33.97 % (Kolamb and Jaya). Within nonbasmati scented rice cultivars, highest genetic dissimilarity of 52.04 % was recorded between Kate chinoor and Kala bhat from Maharashtra, 53.11 % between Mysore sanna and Vasane sanna bhatta from Karnataka, and 49.33 % between Dubraj and Badshahbhog. The dendrogram generated by UPGMA analysis using the distance matrix obtained from AFLP analysis of 53 rice cultivars is shown in Fig. 5.2. At a dissimilarity coefficient of 0.4, the cultivars were grouped into two major clusters (cluster I and cluster II). Cluster I and cluster II comprised 35 and 18 cultivars respectively. Cluster I is subdivided into subcluster IA comprising 30 nonbasmati scented rice cultivars and subcluster IB comprising five basmati cultivars (Fig. 5.2). Subcluster IA characteristically included ambemohar types (Ambemohar, Ambemohar pandhara, Ambemohar Ajra, Ambemohar-Tambda, and Ambemohar-157), along with 25 nonbasmati scented rice landraces.

These 25 rice landraces include ten from Maharashtra (Kala bhat, Kamavatya, Khadkya, Raibhog, Kothmirsal, Kalsal, Jiri, Ghansal, Girga, and Kamod), eight from Karnataka (Gandhesale, Basumati, Kali kumud, Geerige sanna, Medhini sanna bhatta, Vasane sanna bhatta, Kaligajvili, and Kumud), and seven (Kalanamak, Girija sambha, Shyamjeer,

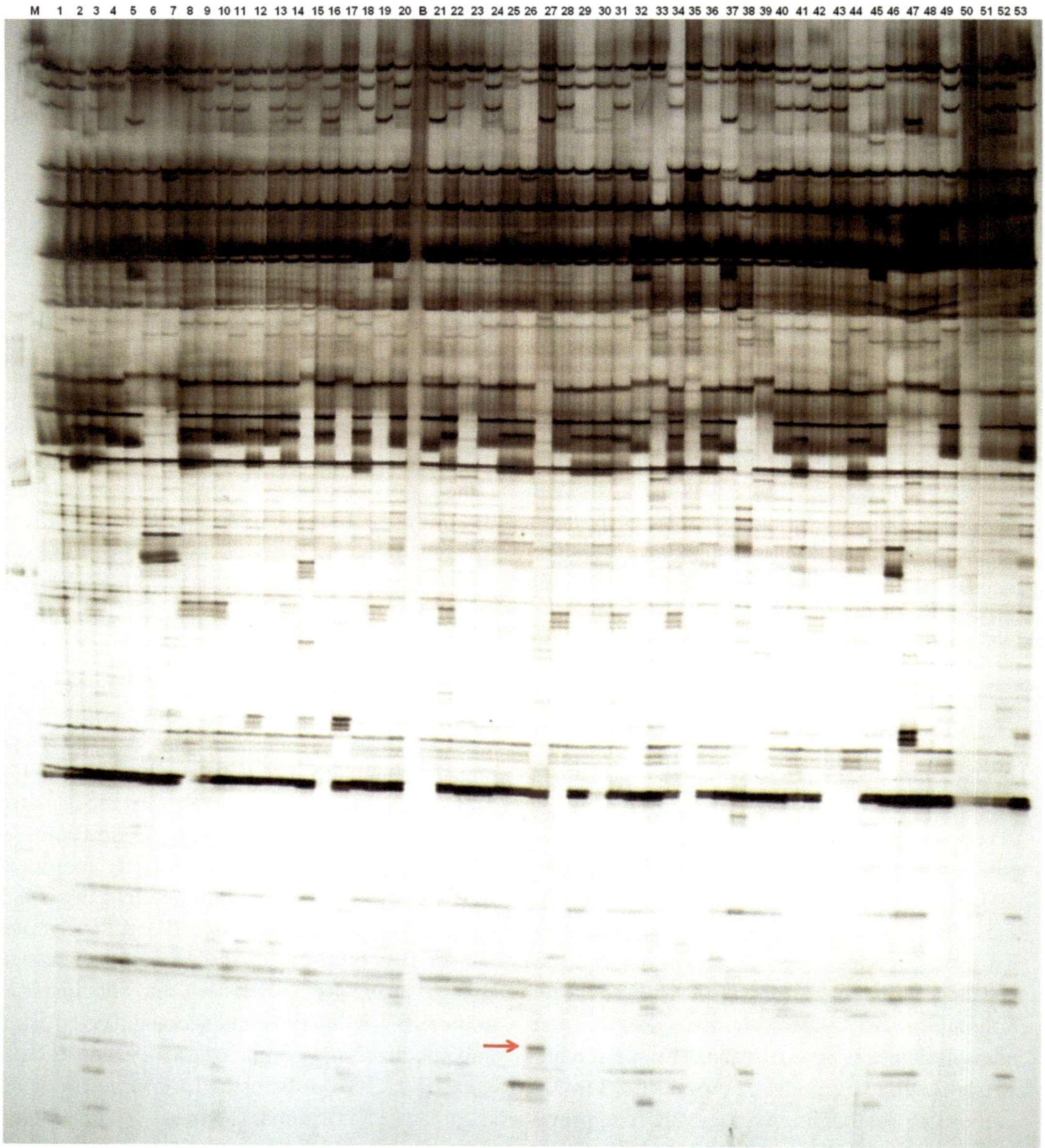

Fig. 5.1 *AFLP profile obtained by primer combination* $E_{ACT}+M_{CAG}$. The *red arrow* indicates the variety specific band for Kagisali. Lane M: ΦX174 DNA (*HaeIII*) Digested DNA 1: Ambemohar, 2: Ambemohar Ajra, 3: Ambemohar Pandhara, 4: Ambemohar-Tambda, 5: Champakali, 6: Chimansal, 7: Gham, 8: Ghansal, 9: Girga, 10: Jiri, 11: Kala bhat, 12: Kalsal, 13: Kamavatya, 14: Kamod, 15: Kate chinoor, 16: Khadkya, 17: Kondhekar chinoor, 18: Kothmirsal, 19: Lal dodki, 20: Raibhog, 21: Velchi, 22: Ambemohar-157, 23: Indrayani, 24: Basumati, 25: Gandhesale, 26: Geerige sanna, 27: Kagisali, 28: Kaligajvili, 29: Kali kumud, 30: Kumud, 31: Medhini sanna bhatta, 32: Mysore sanna, 33: Sanna bili bhatta, 34: Vasane sanna bhatta, 35: Acharmati, 36: Badsahbhog, 37: Bansphool-A, 38: Barke bhat, 39: Dubraj, 40: Dusara, 41: Girija samba, 42: Kalanamak, 43: Shyamjeer, 44: Kalakrishna, 45: Kothmbiri, 46: Kolamb, 47: Manila, 48: Jaya, 49: Kernal local, 50: Pakistan basmati, 51: Basmati 370, 52: Basmati 6311, 53: Pusa basmati-1

Table 5.4 Details of PCR amplified bands generated across 53 rice cultivars by AFLP

Sr No	Primer	Number of scorable bands			% Polymorphc bands	PIC
		Total	Monomorphic	Polymorphic		
1	$E_{ACT}+M_{CAC}$	25	5	20	80.00	0.31
2	$E_{AAC}+M_{CAA}$	31	15	16	51.61	0.28
3	$E_{ACC}+M_{CTG}$	35	10	25	71.43	0.32
4	$E_{ACT}+M_{CAG}$	36	6	30	83.33	0.3
5	$E_{AAG}+M_{CAT}$	49	19	30	61.22	0.35
6	$E_{ACA}+M_{CTA}$	29	4	25	86.21	0.33
7	$E_{AGG}+M_{CTC}$	35	7	28	80.00	0.28
8	$E_{ACG}+M_{CTT}$	27	4	23	85.19	0.31
9	$E_{AGT}+M_{CAC}$	31	12	19	61.29	0.34
10	$E_{AGC}+M_{CAA}$	33	9	24	72.73	0.31
	Total	331	91	240		
	Average	33.1	9.1	24	72.51	0.31

Kalakrishna, Kothimbiri, Badshahbhog, and Dusara) procured from rice research stations.

Dusara separated from the other cultivars indicating higher genetic distance followed by Kaligajvili and Kumud (landraces of Karnataka) within subcluster IA. Basmati type cultivars (Kernal local, Pusa basmati-1, Pakistan basmati, Basmati 370, and Basmati 6311) were grouped as subcluster IB. In this subcluster, Kernal local was placed at farther distance, followed by Pusa basmati-1. Basmati cultivars developed by selection breeding, viz., Basmati 370 and Basmati 6311, were closely placed among basmati type cultivars.

Cluster II included three nonscented cultivars (Kolamb, Manila, and Jaya), nonbasmati scented hybrid cultivar Indrayani, and 14 nonbasmati scented rice landraces. These 15 nonbasmati scented rice cultivars include seven landraces from Maharashtra (Champakali, Lal dodki, Velchi, Chimansal, Kate chinoor, Kondhekar chinoor, and Gham), three from Karnataka (Kagisali, Mysore sanna, and Sanna bili bhatta), and four (Barke bhat, Bansphool-A, Acharmati, and Dubraj) procured from rice research stations. In cluster II, Barke bhat separated from the other cultivars, indicating higher genetic distance.

Mackill et al. (1996) subjected 14 rice accessions (12 *japonica* and two *indica*) to AFLP analysis and obtained average percent polymorphism of 27.7 %, with average of eight polymorphic bands per primer pair combination. They also recorded that almost every AFLP primer combination was usable. All the ten primer combinations used in present study resulted in scorable bands, thus reconfirming the finding of Mackill et al (1996). Zhu et al. (1998) detected 44 % polymorphism among 57 germplasm accessions using four combinations of AFLP primers which generated a total of 410 AFLP bands of which 179 were polymorphic bands, while Bligh et al. (1999) reported only 22 polymorphic bands out of 142 among six long grain rice cultivars. In Egyptian rice cultivars, AFLP amplified 50–90 fragments per primer pair combination and the percentage of polymorphism per primer pair combination is 45.5 % (Saker et al. 2005). Among basmati and elite rice genotypes Aggarwal et al. (2002) obtained 65 % polymorphism while Rashid et al. (2009) recorded that the basmati rice mutants generated an average of 17.63 bands per primer combination and exhibited 37.81 % polymorphism. Prashanth et al. (2002) recorded 62 % polymorphism in cultivars and landraces of *O. sativa* ssp. *indica*. Saini et al. (2004) obtained an average of 34.2 bands per primer combination and 64.0 % average polymorphism using five primer combinations in 18 rice genotypes representing traditional basmati, cross-bred basmati, and nonbasmati rice varieties. In comparison with these reports, the present study recorded higher polymorphism (72.5 %) in the cultivars.

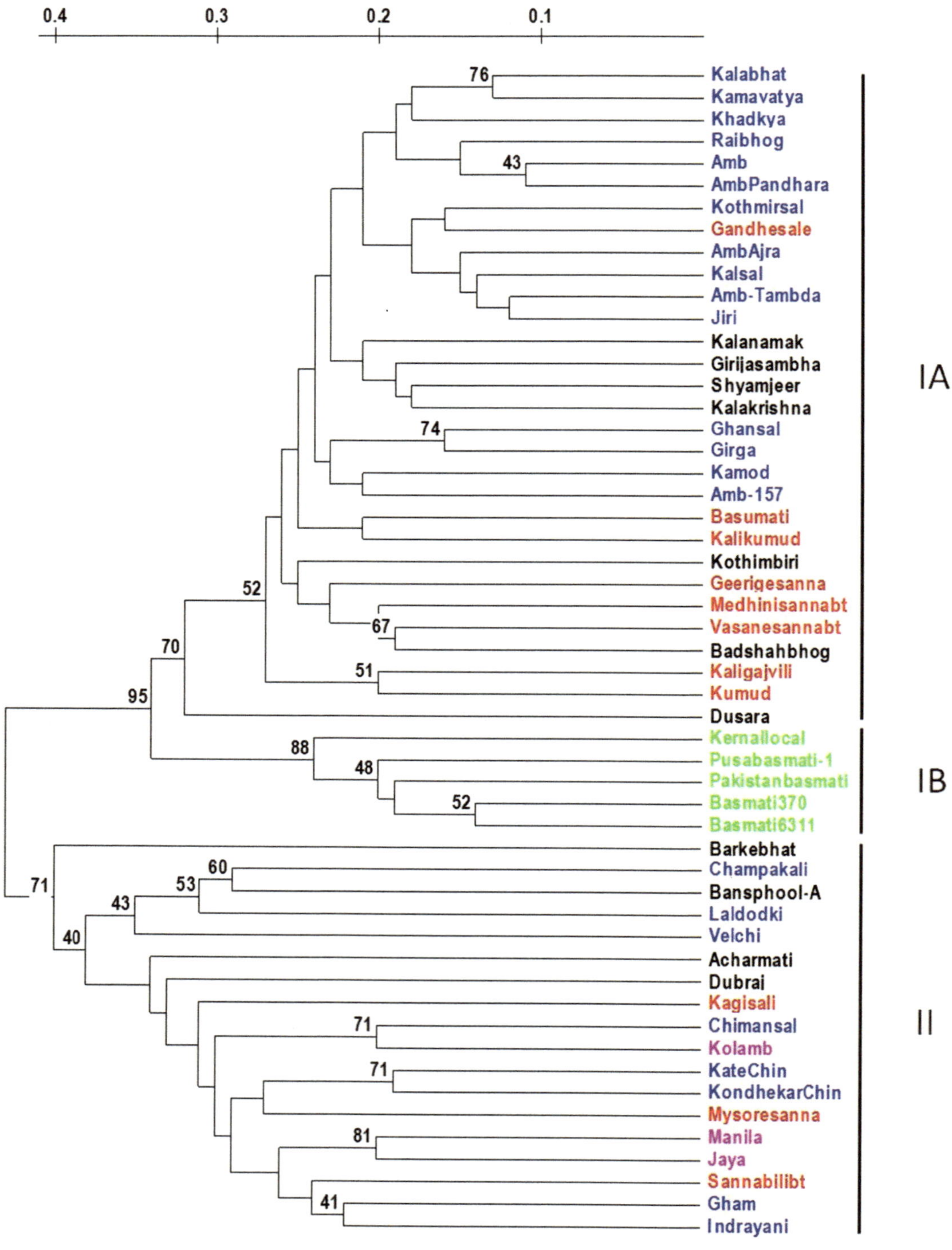

Fig. 5.2 *UPGMA dendrogram showing relationship derived by AFLP among 53 rice cultivars.* (basmati non-basmati scented-Maharashtra, nonbasmati scented-Karnataka, nonbasmati scented – other states of India, non-scented)

A DNA fingerprinting study performed by Singh et al. (1999) revealed 81 % polymorphism among 42 Indian rice varieties by AFLP. Our results indicate higher polymorphism in nonbasmati scented rice cultivars and support this finding. The higher polymorphism detected in the present study could be due to the diversity among the cultivars assessed and the suitability of AFLP markers used in the present study.

Nonbasmati scented rice cultivars exhibited higher diversity in comparison with basmati types studied (Table 5.4 and Fig. 5.2). Among the 53 cultivars assessed, Champakali – a nonbasmati scented landrace from Maharashtra state, India – and Kaligajvili – a nonbasmati scented landrace from Karnataka state, India – recorded highest genetic dissimilarity of 54.21 % (Table 5.4). Within nonbasmati scented rice cultivars from Maharashtra, Kate chinoor grown in Chandrapur district of Maharashtra recorded genetic dissimilarity of 52.04 % from Kala bhat – a landrace grown in selective pockets of Ahmadnagar district of Maharashtra. Similarly, Mysore sanna and Vasane sanna bhatta from Karnataka also recorded high genetic dissimilarity (53.11 %). As seen from the present study, Maharashtra possesses peculiar landraces such as Champakali, Lal dodki, Velchi, Chimansal, Chinoor types (Kate chinoor and Kondhekar chinoor), and Gham, which are genetically diverse from Ambemohar types, Ghansal, and other scented rice cultivars grown within this state. Cultivars from Karnataka, viz., Kagisali, Mysore sanna, and Sanna bili bhatta, were placed in a separate cluster from other nonbasmati scented rice cultivars from Karnataka. Thus, Maharashtra and Karnataka possess genetically diverse rice cultivars. Some of the cultivars included in the present study are native to states other than Maharashtra and Karnataka. Badshahbhog is grown in Bihar, Madhya Pradesh, and West Bengal, Kalanamak in Uttar Pradesh and Bihar. Similarly, Dubraj is grown in Uttar Pradesh and Madhya Pradesh, Giriga sambha in Andhra Pradesh, Acharmati and Kalakrishna in Orrisa, and Shyamjeer in Bihar (Singh et al. 2000a, Rohilla et al. 2000, Shobha Rani and Singh 2003, Das et al. 2003). Among these, Badshabhog and Dubraj exhibited high genetic dissimilarity within nonbasmati scented rice cultivars from other states of India. Singh et al. (2000a) reported that almost every Indian state has its own varieties of scented rice. This is evident from the present study.

Genotype specific segregation was recorded by AFLP analysis irrespective of the cultivar type. Molecular fingerprinting can also suggest if the same variety is cultivated under different names. However, in the present study, no duplication of cultivars was observed. The present study included five cultivars of ambemohar type (Ambemohar, Ambemohar Pandhara, Ambemohar Ajra, Ambemohar-Tambda, and Ambemohar-157), which were segregated in different subclusters in dendrogram. Our study is in agreement with a study by Raghunathachari et al. (2000) on Hansraj accessions by RAPD which revealed that many of the scented varieties under cultivation with similar names are genetically different.

Use of AFLP with ten primer pair combinations in the present study could separate all the basmati type cultivars to form a separate subcluster as supported by higher bootstrap value. Clustering of basmati types into a separate group distinct from that of *indica* and *japonica* rice varieties was reported by using simple sequence length polymorphism (Bligh et al. 1999), fluorescent-AFLP (Aggarwal et al. 2002), fluorescently labeled microsatellite markers (Jain et al. 2004) suggesting that Indian basmati germplasm may have a long, independent, and complex pattern of evolution that distinguishes it from other groups within *Oryza sativa*. Nagaraju et al. (2002) recorded a high degree of genetic similarity among traditional basmati varieties by fluorescence-based ISSR-PCR and SSR markers and indicated that these varieties could be descendants of single landrace and the minor variation observed among them could be as a result of selection practices over years by farmers. Our study supports this finding.

Saini et al. (2004) noted that basmati varieties clustered together with higher degree of similarity irrespective of the marker system and Pusa basmati-1 clustered with basmati group in case of AFLP dendrogram thus support our findings. Mathure et al. (2010) could separate Basmati 370 from Pusa basmati-1 on the basis of combined cluster analysis of RAPD and ISSR; however, they reported that traditional basmati clustered with hybrid basmati. Pusa basmati-1 is a cross between Pusa 150 and Kernal local. Hence, genotypic variation of Pusa basmati-1 might have placed it closer to Kernal local than other basmati cultivars as seen from the dendrogram (Fig. 5.2).

Nonbasmati scented rice cultivars were distributed in cluster I as well as cluster II irrespective of their native state (5.2). ISSR-based study by Joshi et al. (2000) reported that Ambemohar was placed with Khadkya in a subcluster which is separate from Basmati 370 and Indrayani. Among the cultivars under study, the segregation of Kalanamak and Badshahbhog from basmati type has been achieved. Singh et al. (2004) recorded that Kalanamak grouped with Badshahbhog is a separate cluster from Basmati. Similar observations for Kalanamak were recorded by Singh et al. (2006), thus supporting our findings. A diversity analysis among selected nonbasmati scented rice by Mathure et al. (2010) using RAPD and ISSR revealed that Kalanamak, Kamavatya, Raibhog, Ghansal, and Ambemohar clustered together with Basmati types forming a subcluster. They also reported that landrace Gham and hybrid scented cultivar Indrayani were clustered with nonscented rice cultivars, while Lal dodki and Champakali were separated to form another cluster. The present study reported similar clustering pattern. The rice cultivars assessed in the present study vary with respect to agronomic (Mathure et al. 2011) as well as quality traits. Presence of scent in rice is one of the quality traits which represent a very small portion of rice genome. A study by Davierwala et al. (2000) on Indian elite rice varieties by ISSR also reported that nonscented cultivars Jaya and Indrayani were placed closer in a study. Thus supports the clustering of nonscented rice cultivars along with nonbasmati scented rice cultivars as observed in the present study.

The present study provides an insight of genetic diversity among nonbasmati scented rice cultivars from Maharashtra and Karnataka states of India which have largely remained unexplored. Genetically diverse pool of rice germplasm is an important source for the introduction of new quality traits, as well as for broadening genetic base to reduce vulnerability of the crop to epidemics and disasters. The assessment of genetic diversity offers valuable suggestions to plant breeders to identify donor parents and select genetically diverse varieties in crossing programs for crop improvement. In the present study, no duplicates are observed among the rice cultivars; thus, AFLP markers could segregate closely related varieties such as ambemohar types with higher resolution. This genetic variance allows the selection of varieties with important agronomic and quality traits that can be utilized in plant breeding. The study revealed the genetic variability among nonbasmati scented rice cultivars which further necessitates protection and conservation of these cultivars as genetic resources.

5.3.1 Genotype Specific Bands

AFLP primer combinations among 53 cultivars yielded 12 genotype specific bands. Details of the primer combination, genotype, and approximate band size are given in Table 5.5. Identification of different genotypes is essential for characterization of diverse accessions of germplasm. One of the most important facets of molecular diversity studies is to identify markers that can discriminate between different varieties. AFLP is reported as efficient and useful to generate genotype-specific fingerprints in Indian elite and basmati rice (Singh et al. 1999; Aggarwal et al. 2002; Jain et al. 2004), Cuban rice (Fuentes et al. 1999), and Egyptian rice (Saker et al. 2005) cultivars.

As seen from Table 5.5, nonbasmati scented cultivars, especially landraces, amplified genotype-specific bands. DNA fingerprinting of Indian rice by Singh et al. (1999) also reported that among the cultivars assessed, landraces exhibit distinct identity and amplified unique loci. Jain et al. (2004) recorded that landraces such as Ambemohar from Maharashtra and aromatic short grain traditional varieties from Assam are genetically distinct and revealed a higher number of unique alleles. Saker et al. (2005) could detect 65 unique AFLP markers in seven Egyptian rice cultivars. The present study resulted

Table 5.5 Distance matrix generated by AFLP analysis indicating interrelation between 53 rice cultivars

Sr	Cultivar	1	2	3	4	5	6	7	8	9	10	11	12	13
1	Amb	0												
2	AmbAjra	17.21	0											
3	AmbPandhara	11.28	15.29	0										
4	Amb-Tambda	17.21	14.05	13.73	0									
5	Champakali	44.21	47.19	40.98	40.26	0								
6	Chimansal	38.98	41.88	39.27	40.17	37.22	0							
7	Gham	37.35	41.7	35.39	41.7	32.2	23.85	0						
8	Ghansal	25	21.01	24.3	26.89	46.26	40.87	39.92	0					
9	Girga	19.3	22.67	21.19	25.33	44.86	37.04	40.35	15.74	0				
10	Jiri	15.79	15.92	11.63	11.84	47.01	41.77	44	25.31	22.47	0			
11	Kalabhat	18.61	22.27	19.84	16.16	47.71	44.8	47.86	29.78	22.86	18.1	0		
12	Kalsal	19.18	16.05	17.19	11.93	41.38	42.98	41.94	23.85	25.33	16.26	17.39	0	
13	Kamavatya	19.34	21.16	20.47	20.33	44.35	40.77	40.65	22.36	22.73	21.31	13.16	16.53	0
14	Kamod	21.99	23.85	21.43	25.52	47.37	42.86	42.62	25.11	22.73	23.14	22.12	22.5	19.33
15	KateChin	42.37	47.01	40.89	45.3	42.6	23.01	23.85	46.96	46.05	45.15	52.04	43.83	44.21
16	Khadkya	22.22	25.86	22.45	23.28	42.99	45.54	40.93	20.18	27.7	26.81	19.64	21.03	16.88
17	KondhekarChin	43.5	46.61	41.03	46.61	45.71	27.7	28.32	42.86	40.59	45.54	50	46.85	41.82
18	Kothmirsal	25.52	18.99	21.6	19.83	50.44	44.98	44.63	27.04	27.85	18.33	22.32	15.13	23.73
19	Laldodki	39.24	39.57	33.87	37.02	33.04	35.68	35.83	39.39	37.96	36.13	40.54	38.14	38.46
20	Raibhog	13.56	23.08	16.6	20.51	47.09	45.13	43.1	25.22	23.36	21.52	15.84	21.7	17.6
21	Velchi	34.21	42.48	36.4	37.17	37.67	39.45	34.2	35.14	37.44	39.74	39.91	39.21	35.11
22	Amb-157	20.69	27.83	20.17	26.96	47.03	41.44	42.13	23.01	19.43	21.03	24.42	23.81	22.27
23	Indrayani	35.09	38.94	33.89	37.17	43.26	30.28	22.08	39.64	36.84	37.12	40.85	37.45	35.11
24	Basumati	24.59	21.49	20.78	22.31	48.05	45.3	43.32	29.41	28.83	20.82	24.02	20.17	26.97
25	Gandhesale	24.28	17.01	20.47	18.67	47.83	41.63	41.46	22.36	24.89	16.39	22.81	18.18	25
26	Geerigesanna	28.76	23.81	27.87	28.14	49.09	41.7	44.07	24.23	24.53	25.64	28.44	28.45	28.7
27	Kagisali	31.75	34.63	32.42	33.66	43	29	30.19	33.98	33.33	32.38	37.76	36.84	33.33

(continued)

Table 5.5 (continued)

Sr	Cultivar	1	2	3	4	5	6	7	8	9	10	11	12	13
28	Kaligajvili	26.87	29.78	26.89	27.11	54.21	45.62	44.35	24.89	28.5	25.44	29.25	30.09	27.68
29	Kalikumud	27.43	21.43	21.52	24.11	49.3	45.37	43.23	28.18	30.77	22.47	29.86	21.78	30.94
30	Kumud	28.89	26.46	27.12	30.05	49.06	43.26	42.98	19.64	26.47	28.32	32.38	30.36	26.13
31	Medhinisannabt	20.17	23.14	20	25.44	41.55	37.9	35.65	21.97	20.18	21.55	24.42	24.35	23.01
32	Mysoresanna	44.4	47.51	41.88	46.61	42.86	29.58	30.97	47.47	42.57	43.75	49.04	45.05	45.46
33	Sannabilibt	36.44	37.61	33.6	38.46	41.7	30.97	23.85	37.39	41.12	38.4	42.08	37.02	36.48
34	Vasanesannabt	24.14	26.09	23.46	27.83	47.03	45.95	47.23	26.55	25.47	24.46	28.11	26.41	28.38
35	Acharmati	40.81	42.99	41.88	43.89	43.81	34.27	32.74	34.56	40.59	42.86	45.19	43.24	37.27
36	Badshahbhog	25.11	22.75	25.2	28.76	47.75	45.78	45.38	24.89	27.44	27.12	30.91	28.21	27.59
37	Bansphool-A	38.27	37.76	33.07	35.27	28.7	38.2	33.33	36.71	39.37	37.71	42.11	38.02	40
38	Barkebhat	40.86	45.95	41.54	43.62	40.91	38.98	39.27	45.06	38.29	44.5	45.25	43.32	40.86
39	Dubraj	43.22	41.88	39.27	41.03	41.7	32.74	28.87	39.13	45.54	44.3	44.8	41.28	42.49
40	Dusara	35.61	27.36	31.13	27	53.13	50	49.75	34.36	37.9	27.45	33.33	30.69	30.3
41	Girijasambha	23.48	18.42	23.65	23.68	50.23	46.36	45.06	23.21	26.07	18.61	26.51	24.02	25.11
42	Kalanamak	23.4	26.18	23.58	25.32	50.45	46.67	45.38	26.64	25.82	22.88	22.73	26.5	27.59
43	Shyamjeer	20.21	23.66	20.21	21.98	53.67	46.07	44.5	22.83	25.42	19.15	20	26.88	22.58
44	Kalakrishna	23.42	20	24.46	25.46	51.2	48.11	46.67	22.22	29.06	25.56	29.47	23.08	26.03
45	Kothimbiri	21.01	28.81	23.7	23.73	51.11	42.98	44.4	29.31	29.03	23.85	23.77	26.58	25.11
46	Kolamb	45.13	46.43	44.3	47.32	40.85	20.37	26.64	41.82	42.72	46.26	51.66	45.78	44.4
47	Manila	39.92	41.08	35.43	40.25	36.52	29.61	25.2	41.77	39.37	42.62	44.74	39.67	36.67
48	Jaya	42.36	43.61	39.17	40.09	37.96	30.59	25	42.6	43.96	40.87	42.06	38.6	38.05
49	Kernallocal	27.43	29.46	30.8	32.14	49.3	50.93	45.85	27.27	30.39	30.4	30.81	29.78	28.25
50	Pakistanbasmati	35.67	39.76	34.09	35.33	49.37	40.88	41.71	39.29	39.87	35.67	32.1	35.33	36.47
51	Basmati370	26.9	27.08	25.49	28.13	48.91	47.59	44.28	31.28	37.08	28.93	26.88	25.77	29.59
52	Basmati6311	37.74	38.46	36.36	35.27	52.76	44	45.37	40.1	42.41	37.44	33.67	35.89	37.5
53	Pusabasmati-1	31.97	31.41	27.84	33.88	46.32	39.32	37.65	36.98	41.18	34.69	37.12	30.04	33.61

Sr	Cultivar	14	15	16	17	18	19	20	21	22	23	24	25	26
15	KateChin	41.13	0											
16	Khadkya	22.27	50	0										
17	KondhekarChin	41.28	19.25	45.97	0									
18	Kothmirsal	26.5	47.6	25.11	50	0								
19	Laldodki	39.66	36.56	42.22	34.58	45.22	0							
20	Raibhog	18.61	46.9	18.75	48.36	24.02	41.85	0						
21	Velchi	35.43	42.2	34.26	35.61	46.61	32.42	36.7	0					
22	Amb-157	20.71	43.24	23.64	43.54	25.33	36.32	18.92	40.19	0				
23	Indrayani	37.22	23.85	43.52	27.81	39.37	38.81	38.53	37.14	39.25	0			
24	Basumati	27.2	47.86	29.31	50.23	21.52	42.13	24.79	46.9	28.7	41.59	0		
25	Gandhesale	22.69	45.06	29	45.46	16.1	38.46	22.75	41.33	31	36.89	19.5	0	
26	Geerigesanna	28.95	46.19	33.03	44.76	27.43	42.86	26.46	43.26	26.94	42.33	30.74	19.13	0
27	Kagisali	35.27	33.01	36.95	26.7	39.6	34.65	35.32	35	34.65	29.53	38.92	36.95	37.37
28	Kaligajvili	27.03	43.78	24.65	45.1	27.27	47.71	20.74	36.84	26.76	38.76	30.67	29.46	32.71
29	Kalikumud	26.7	46.3	30.84	50.74	16.9	44.7	26.85	44.23	28.3	41.35	21.43	19.28	28.64
30	Kumud	23.64	45.12	25.82	46.54	30.28	45.37	25.58	35.27	27.01	39.13	33.63	27.03	25.47
31	Medhinisannabt	25.11	39.37	27.6	39.13	25	32.73	21.82	37.62	21.82	32.39	28.63	20.89	23.85
32	Mysoresanna	46.79	26.76	50.71	27	50	34.58	48.36	44.39	43.54	30.73	47.51	44.55	47.62
33	Sannabilibt	36.8	29.2	40.18	33.33	38.87	37.45	39.82	39.45	41.44	23.85	37.61	35.62	44.4
34	Vasanesannabt	26.87	48.65	29.09	50.24	27.11	41.7	20.72	42.99	23.85	45.79	33.91	26.64	22.37
35	Acharmati	36.7	37.09	41.23	31	45.37	41.12	44.6	34.63	42.58	35.61	48.42	39.09	45.71
36	Badshahbhog	24.35	44	28.25	41.51	28.95	43.36	24.44	40.09	24.89	43.78	32.19	25	22.52
37	Bansphool-A	40.34	39.06	37.66	42.73	42.37	29.06	38.2	36	37.99	39.56	40.25	40	40
38	Barkebhat	43.48	39.77	47.49	37.21	51.35	45.36	44.75	42.7	44	37.57	49.46	44.97	42.05
39	Dubraj	45.46	31.86	41.07	35.21	45.85	40.09	45.13	44.04	45.05	35.78	41.88	39.91	47.09
40	Dusara	35.68	50.52	37.82	48.89	31.63	46.11	33.33	46.32	36.46	40.86	33.67	27.92	31.58
41	Girijasambha	26.22	47.27	29.36	47.83	24.66	40.27	23.64	39.62	25.93	40.57	27.19	17.18	18.89
42	Kalanamak	29.57	49.33	29.15	52.83	25.44	42.48	22.67	42.86	25.79	44.7	25.32	24.14	24.32
43	Shyamjeer	25.41	49.44	19.13	46.11	24.18	44.32	19.55	33.33	28.09	41.86	30.06	24.59	31.82

(continued)

Table 5.5 (continued)

Sr	Cultivar	14	15	16	17	18	19	20	21	22	23	24	25	26
44	Kalakrishna	26.27	50	22.86	47.74	24.65	47.42	20.76	37.26	30.77	42.16	30.91	20.55	24.4
45	Kothimbiri	27.04	42.98	30.09	44.19	27.27	39.74	19.3	42.73	24.11	39.09	29.66	24.26	26.22
46	Kolamb	46.61	31.48	48.6	33.99	43.38	37.33	49.07	39.42	43.4	31.73	47.32	41.7	45.54
47	Manila	38.66	27.04	40.26	27.27	42.37	26.5	44.21	40.44	39.74	29.78	37.76	40	45.22
48	Jaya	40.18	26.94	39.17	30.1	44.14	29.09	42.47	39.34	41.4	25.12	39.21	38.94	46.3
49	Kernallocal	28.51	50	28.04	47.78	29.68	46.54	24.07	41.35	31.13	44.23	26.79	29.15	32.39
50	Pakistanbasmati	36.84	43.21	37.04	38.07	39.02	41.82	33.33	36.53	37.89	31.65	33.33	34.5	42.33
51	Basmati370	30.96	46.32	30.16	46.37	29.47	43.92	24.35	40.43	35.11	39.13	21.24	26.15	33.69
52	Basmati6311	36.23	43.28	37.31	43.46	33.66	46.54	32.67	43.15	42	39.49	32.38	33.65	41.71
53	Pusabasmati-1	34.73	39.32	37.93	42.08	33.33	43.83	33.33	45.13	37.39	36.28	30.58	32.78	39.39
Sr	Cultivar	27	28	29	30	31	32	33	34	35	36	37	38	39
28	Kaligajvili	33.33	0											
29	Kalikumud	39.58	25.6	0										
30	Kumud	34.72	20.39	21.95	0									
31	Medhinisannabt	33.01	27.36	25.47	24.17	0								
32	Mysoresanna	32.62	50.98	48.77	45.55	38.84	0							
33	Sannabilibt	30.61	41.01	37.96	37.67	38.6	28.64	0						
34	Vasanesannabt	37.69	29.58	29.25	31.75	20.18	53.11	50.45	0					
35	Acharmati	35.42	40.2	43.84	35.64	41.75	34	27.7	51.2	0				
36	Badshahbhog	35.64	30.56	29.3	27.1	20.72	50	44.89	18.55	46.23	0			
37	Bansphool-A	34.63	40.18	44.4	36.94	34.51	37.27	38.2	41.49	42.73	37.93	0		
38	Barkebhat	35.29	45.46	50	47.62	36.42	34.52	39.33	43.18	40.23	38.64	40.86	0	
39	Dubraj	39.7	43.78	43.52	43.26	42.47	35.21	29.2	53.15	35.21	49.33	35.62	41.24	0
40	Dusara	39.23	30.81	30.81	33.7	33.33	50.84	37.57	38.95	44.44	35.05	45.46	43.42	40.63

41	Girijasambha	37.76	25.12	21.91	23.45	23.72	46.86	34.55	25.93	37.2	23.29	42.73	41.71	43.64
42	Kalanamak	40.1	26.85	26.51	28.04	21.46	50.94	39.56	25.79	47.17	26.79	41.38	45.46	41.33
43	Shyamjeer	40.57	19.77	24.42	24.42	27.75	51.81	39.08	29.21	39.53	28.18	45.06	43.05	41.24
44	Kalakrishna	38.3	22.17	25.74	23.38	24.04	50.75	43.4	25	42.71	22.28	43.38	45.78	44.34
45	Kothimbiri	37.26	25.11	28.44	30.88	21.27	44.19	42.11	27.68	43.26	25.11	42.98	43.78	46.49
46	Kolamb	37.5	47.83	47.57	47.32	37.14	32.02	33.33	50	33.01	46.05	43.5	40	35.19
47	Manila	31.1	47.32	41.7	44.14	35.71	30.91	26.18	47.6	37.27	44.83	37.5	41.3	29.61
48	Jaya	30.26	43.81	43.54	44.23	38.39	32.04	25.11	51.63	38.84	48.62	38.05	43.18	28.77
49	Kernallocal	44.79	33.33	33.01	35.61	29.52	50.74	44.44	30.19	49.75	31.16	46.19	46.51	47.22
50	Pakistanbasmati	41.72	41.98	40.65	43.59	40.74	37.84	36.2	47.24	40.13	41.82	44.71	38.46	35.37
51	Basmati370	42.86	35.83	29.67	37.36	29.84	45.46	38.3	34.39	48.07	35.42	45.18	41.56	38.22
52	Basmati6311	41.11	36.41	37.17	41.36	35	46.52	43.28	44.72	49.74	42.57	49.28	46.34	41.58
53	Pusabasmati-1	39.51	40.44	32.14	37.22	32.74	37.56	34.19	42.61	42.08	40.77	41.08	46.07	33.33

Sr	Cultivar	40	41	42	43	44	45	46	47	48	49	50	51	52
41	Girijasambha	19.15	0											
42	Kalanamak	26.7	18.72	0										
43	Shyamjeer	32.1	18.86	18.39	0									
44	Kalakrishna	33.33	18.45	25.12	17.65	0								
45	Kothimbiri	32.99	24.32	25.11	25.97	27.1	0							
46	Kolamb	51.09	46.67	51.63	48.84	47.53	45.87	0						
47	Manila	46.19	45.37	43.97	45.86	48.86	42.98	27.35	0					
48	Jaya	41.3	42.72	45.87	42.35	47.32	42.08	33.97	20.35	0				
49	Kernallocal	37.36	25.71	25.58	27.49	29.7	30.28	49.52	42.6	41.63	0			
50	Pakistanbasmati	32.87	33.74	35.37	37.31	38.16	37.28	38.07	36.15	32.91	23.57	0		
51	Basmati370	32.93	27.96	27.66	29.94	28.81	29.9	45.86	37.44	36.96	16.2	19.23	0	
52	Basmati6311	37.57	35.35	34.65	37.04	36.84	36.95	41.97	39.51	36.79	25.13	18.71	14.45	0
53	Pusabasmati-1	32.66	35.97	35.62	40.98	35.46	38.14	35.71	31.95	33.92	30.36	19.51	17.53	22.33

in fewer number of genotype-specific bands as compared to Saker et al. (2005); this could be due to the higher number of cultivars under study. Joshi et al. (2000) and Aggarwal et al. (2002) highlighted the need to assess specificity of unique bands over a large sample size. This was addressed in the present study to ensure utility of unique bands even in closely related cultivars.

DNA fingerprint is a useful tool for varietal identification, authentication of genotypes, generating mapping populations, and DUS (distinctness, uniformity and stability) testing for parents and plant breeder's rights. The genotype specific bands obtained in the present study can be further exploited as a diagnostic for cultivar identification.

5.4 Validation of Aroma Allele

To validate the marker designed by Amarawathi et al. (2008), 54 cultivars were subjected to validation of marker for aroma allele by Mathure et al (2014). These included 46 nonbasmati scented, five basmati, and three nonscented cultivars. All basmati cultivars amplified band of 82 bp and nonscented cultivars amplified 90 bp band, thus complying with the results indicated by Amarawathi et al (2008). Among nonbasmati scented types, 82 bp band was amplified in 26 cultivars, 16 cultivars exhibited banding pattern of nonscented cultivar (90 bp band), and four cultivars, viz., Ambemohar Pandhara, Ambemohar-157, Basumati, and Khadkya, amplified both bands (indicated by black arrows in Fig. 5.3). These 16 nonbasmati scented rice cultivars exhibiting banding pattern of nonscented trait included Acharmati, Ambemohar-Tambda, Banasphool-A, Barke bhat, Champakali, Chimansal, Gham, Jiri, Kagisali, Kalanamak, Kalsal, Lal dodki, Mysore sanna, Sanna bili bhatta, Tamsal, and Velchi.

This study included Ambemohar, Ambemohar Ajra, Ambemohar Pandhara, Ambemohar-Tambda, and Ambemohar-157. Among these, 82 bp band was amplified by Ambemohar Ajra, 90 bp band by Ambemohar and Ambemohar-Tambda, while Ambemohar Pandhara, Ambemohar-157 were heterozygous. Thus, Ambemohar cultivars are popularly cultivated in the Maval tract of Maharashtra, India, and exhibit variation with respect to aroma locus (Mathure et al. 2014). A study by Sakthivel et al. (2009) reported five indigenous short grain scented landraces, including Banasphool-A, which gave nonfragrant banding pattern due to absence of 8 bp deletion. Similar finding was recorded in for Banasphool-A. Absence of 8 bp deletion was also recorded in a number of scented varieties primarily from South and Southeast Asia (Fitzgerald et al. 2008). Such exceptions were also recorded in 12 rice varieties by Shi et al. (2008) and in Gobindabhog and Tulsibhog by Roy et al. (2012).

Among scented rice cultivars without the 8 bp deletion in exon 7 of *badh2* gene, Champakali, Chimansal, Velchi, and Kagisali recorded 2AP content of 0.367, 0.292, 0.224, and 0.249 mg/kg respectively (Mathure et al. 2014). Presence of 2AP in these cultivars could be as a result of some other mutation in *badh2* gene. Several rare mutations in different regions of *badh2* gene such as in promoter (Bourgis et al. 2008), exon 2 (Shi et al. 2008), exon 8 (Amarawathi et al. 2008, Sun et al. 2008), exon 13 (Myint et al. 2012), and introns 2 and 4 (Chen et al. 2008), besides 8 bp deletion, are reported in rice. Further study of these cultivars with respect to DNA sequence can lead to the identification of new alleles of *badh2* associated with 2AP synthesis in these cultivars.

Mathure et al (2014) reported positive correlation of 2AP content with the presence of 82 bp band and negative correlation with the presence of 90 bp band using Pearson's correlation coefficients for band size and 2AP content. Though the detailed biosynthesis pathway of 2AP has not yet been completely elucidated (Lorieux et al. 1996; Bradbury et al. 2005a), positive correlation between 2AP content and presence of 82 bp band attests the role of 8 bp deletion causing inactive *badh2* in the accumulation of 2AP in rice.

In 2009, Sakthivel et al. (2009) analyzed basmati, indigenous short grain scented varieties, long grain scented, and nonscented rice varieties by a simple functional marker based on 8 bp deletion in the exon 7 of *badh2* and suggested its use in rice breeding programs. This marker allowed segregation of basmati from nonscented cultivars and identification of heterozygotes. However, 16 nonbasmati scented cultivars exhib-

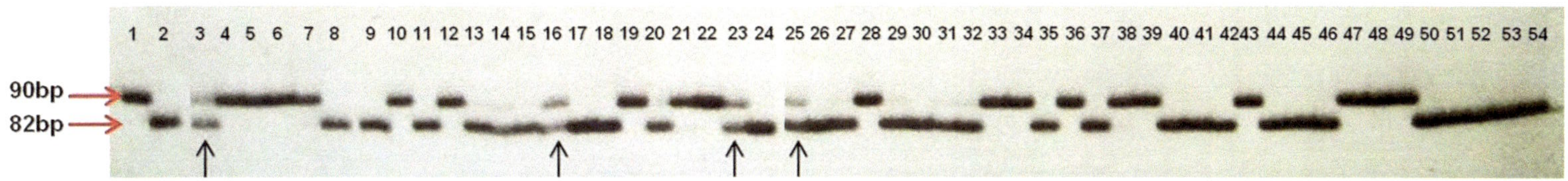

Lane 1:Ambemohar, 2:Ambemohar Ajra, 3:Ambemohar Pandhara, 4:Ambemohar-Tambda, 5:Champakali, 6:Chimansal, 7:Gham, 8:Ghansal, 9:Girga, 10:Jiri, 11:Kala bhat, 12:Kalsal, 13:Kamavatya,14: Kamod, 15:Kate chinoor, 16:Khadkya, 17:Kondhekar chinoor, 18:Kothmirsal, 19:Lal dodki, 20:Raibhog, 21:Tamsal, 22:Velchi, 23:Ambemohar-157, 24:Indrayani, 25:Basumati, 26:Gandhasale, 27:Geerige sanna, 28:Kagesali, 29:Kaligajvili, 30:Kali kumud, 31:Kumud, 32:Medhini sanna bhata, 33:Mysore sanna, 34:Sanna bili bhata, 35:Vasane sanna bhata, 36:Acharmati, 37:Badsahbhog, 38:Bansaphool-A, 39:Barke bhat, 40:Dubraj, 41:Dusara, 42:Girija samba, 43:Kalanamak, 44:Shyamjeer, 45:Kalakrishna, 46:Kothmbiri, 47:Kolamb, 48:Manila, 49:Jaya, 50:Kernal local, 51:Pakistan basmati, 52:Basmati 370, 53:Basmati 6311, 54:Pusa basmati-1

Fig. 5.3 *PAGE gel indicating amplification of aroma allele(s) among 54 rice cultivars. Vertical arrows* indicate heterozygous bands. Lane 1: Ambemohar, 2: Ambemohar Fig. 5.3 (continued) Ajra, 3: Ambemohar Pandhara, 4: Ambemohar-Tambda, 5: Champakali, 6: Chimansal, 7: Gham, 8: Ghansal, 9: Girga, 10: Jiri, 11: Kala bhat, 12: Kalsal, 13: Kamavatya, 14: Kamod, 15: Kate chinoor, 16: Khadkya, 17: Kondhekar chinoor, 18: Kothmirsal, 19: Lal dodki, 20: Raibhog, 21: Tamsal, 22: Velchi, 23: Ambemohar-157, 24: Indrayani, 25: Basumati, 26: Gandhasale, 27: Geerige sanna, 28: Kagesali, 29: Kaligajvili, 30: Kali kumud, 31: Kumud, 32: Medhini sanna bhata, 33: Mysore sanna, 34: Sanna bili bhata, 35: Vasane sanna bhata, 36: Acharmati, 37: Badsahbhog, 38: Bansaphool-A, 39: Barke bhat, 40: Dubraj, 41: Dusara, 42: Girija samba, 43: Kalanamak, 44: Shyamjeer, 45: Kalakrishna, 46: Kothmbiri, 47: Kolamb, 48: Manila, 49: Jaya, 50: Kernal local, 51: Pakistan basmati, 52: Basmati 370, 53: Basmati 6311, 54: Pusa basmati-1

ited nonscented banding pattern rendering the marker as unsuitable for MAS of nonbasmati scented rice (Mathure et al. 2014).

5.4.1 Validation of *Wx*-SSR for Amylose Content

Fifty-four Indian rice cultivars were subjected to PCR amplification of SSR linked to *Wx* gene. Separation on PCR products on native PAGE revealed the presence of six alleles among the cultivars assessed. Allele 2 and allele 4 were more frequent and detected in 24 and 15 cultivars respectively. Alleles 1, 3, 5, and 6 were less frequent and were present in 3, 3, 8, and 1 cultivars respectively (Fig. 5.4). The number of alleles detected for *Wx*-SSR varies from as less as four (Bligh et al. 1995; Bao et al. 2002) to ten (Prathepha 2003; Bao et al. 2006). These alleles are reported to vary in size from 102 to 132 bp (Ayres et al. 1997; Bao et al. 2006; Rahemi et al. 2007). The size of products observed (~80 to ~140 bp) in the present study is comparable with the earlier reports.

Uneven distribution of microsatellite alleles in population resulting in higher frequency of two alleles was reported by Shu et al. (1999). They reported that alleles with 11 and 18 (CT) repeats were frequent *indica* while 16 and 17 repeats were frequent in *japonica* rice (Shu et al. 1999). Similarly, allele with eight repeats in Iranian rice (Rahemi et al. 2007), 17 repeats in Thai rice (Prathepha 2003; Prathepha and Baimai 2004) and in waxy rice (Bao et al. 2002), and 18 repeats in Chinese rice (Bao et al. 2006) was present in higher frequency.

Nonbasmati cultivars exhibited higher variation in the microsatellite assessed as all six alleles of *Wx*-SSR were detected in one or more nonbasmati scented rice cultivars. Basmati cultivars amplified allele 2 (Kernal local, Basmati 370, Basmati 6311) and allele 5 (Pakistan basmati and Pusa basmati-1). Alleles 4 and 5 were amplified in nonscented rice cultivars. This could be due to diversity in genetic makeup and higher number of nonbasmati scented cultivars analyzed in the present study. Bao et al. (2006) recorded that Chinese landraces exhibited unique alleles and differ from breeding lines with respect to size of microsatellite allele. In the present study, allele 6 with lowest molecular weight was uniquely amplified in Mysore sanna, a landrace (indicated by black arrow in Fig. 5.4), thus supporting their finding.

Among five ambemohar types (Ambemohar, Ambemohar Ajra, Ambemohar Pandhara, Ambemohar-Tambda and Ambemohar-157) analyzed, three alleles for *Wx*-SSR were detected. Allele 1 was present in Ambemohar and Ambemohar Ajra, allele 2 in Ambemohar-Tambda, and allele 4 in Ambemohar Pandhara and Ambemohar 157. The variation exhibited in ambemohar types with respect to *Wx*-SSR will help in further segregation of Ambemohar with same aroma locus.

Average polymorphism information content (PIC) of cultivars varied as 0.25 in nonbasmati scented rice, 0.48 in basmati, and 0.44 in nonscented rice cultivars. Bao et al. (2006) reported PIC of 0.224 in landraces and 0.649 for breeding lines. These values are in agreement with the present study.

Average apparent amylose content observed for each allele of *Wx*-SSR is presented in Table 5.7. Cultivars with alleles 1, 2, and 3 had significantly lower apparent amylose content than Mysore sanna with allele 6 as revealed by Duncan's multiple range test. Thus, alleles with higher molecular weight corresponded with lower apparent amylose content. Similar finding was reported by Ayres et al. (1997), Shu et al. (1999), Bergman et al. (2001) and Bao et al. (2006). Ayres et al. (1997) reported that the cultivars in the CT repeat classes with n 18, 14 or 20, and 11 contained 14–19 %, 20–23 %, and greater than 23 % amylose respectively, while cultivars with (CT)17 and (CT)19 repeats exhibited divergent amylose contents. Bergman et al. (2001) reported that the CT repeat classes with n (10 or 11, 14 or 20, and 17 or 18) were categorized as high-amylose, intermediate-amylose, and low-amylose types, respectively. Bao et al. (2006) recorded that apparent amylose content of the allele classes with lower number of CT repeats (8, 10, 11, and 12 repeats) was significantly higher than other classes with higher number of CT repeats (17, 18, and 19 repeats).

Pearson's correlation coefficients between presence of allele and apparent amylose content were calculated. Presence of allele 2 correlated

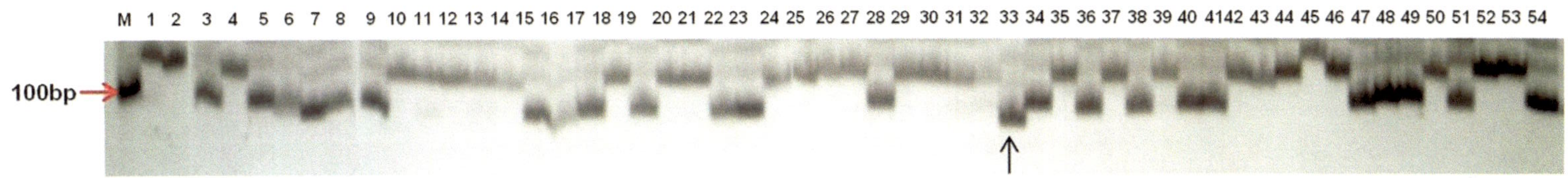

Fig. 5.4 *PAGE gel indicating PCR products of Wx-SSR among 54 rice cultivars. Black arrow* indicates unique allele (allele 6) amplified in Mysore sanna. 1: Ambemohar, 2: Ambemohar Ajra, 3: Ambemohar Pandhara, 4: Ambemohar-Tambda, 5: Champakali, 6: Chimansal, 7: Gham, 8: Ghansal, 9: Girga, 10: Jiri, 11: Kala bhat, 12: Kalsal, 13: Kamavatya, 14: Kamod, 15: Kate chinoor, 16: Khadkya, 17: Kondhekar chinoor, 18: Kothmirsal, 19: Lal dodki, 20: Raibhog, 21: Tamsal, 22: Velchi, 23: Ambemohar-157, 24: Indrayani, 25: Basumati, 26: Gandhasale, 27: Geerige sanna, 28: Kagesali, 29: Kaligajvili, 30: Kali kumud, 31: Kumud, 32: Medhini sanna bhata, 33: Mysore sanna, 34: Sanna bili bhata, 35: Vasane sanna bhata, 36: Acharmati, 37: Badsahbhog, 38: Banasphul A, 39: Barke bhat, 40: Dubraj, 41: Dusara, 42: Girija samba, 43: Kalanamak, 44: Shyamjeer, 45: Kalakrishna, 46: Kothmbiri, 47: Kolamb, 48: Manila, 49: Jaya, 50: Kernal local, 51: Pakistan basmati, 52: Basmati 370, 53: Basmati 6311, 54: Pusa basmati-1

Table 5.6 Details of genotype specific bands generated during AFLP analysis

Sr No	Primer combination	Genotype specific bands
1	$E_{ACT}+M_{CAC}$	Mysore sanna (~550 bp)
		Banasphool-A (~280 bp)
2	$E_{AAC}+M_{CAA}$	Dubraj (~220 bp)
3	$E_{ACC}+M_{CTG}$	Geerige sanna (~80 bp)
4	$E_{ACT}+M_{CAG}$	Kagisali (~90 bp)
5	$E_{ACA}+M_{CTA}$	Sanna bili bhatta (~280 bp)
		Geerige sanna (~75 bp)
6	$E_{AGG}+M_{CTC}$	Jaya(~200 bp)
		Champakali (~120 bp)
		Dusara (~75 bp)
7	$E_{AGT}+M_{CAC}$	Jaya (~190 bp)
		Kothimbiri (~170 bp)

Table 5.7 Details of microsatellite alleles and average apparent amylose content for each allele among 54 Indian rice cultivars

Allele number	Molecular weight of allele	Average apparent amylose content
1	High	22.88^{a}
2	↓	22.14^{a}
3		23.81^{a}
4		25.39^{ab}
5	Low	25.33^{ab}
6		29.14^{b}

Values with the same letter superscripted in each allele number are not significantly different at $p=0.05$

with reduction in amylose content ($r=-0.449$) at 0.01 significance level, while allele 4 correlated with increase in amylose content ($r=0.306$) at 0.05 significance level. Such associations were also recorded in US (Ayres et al. 1997) and Thai (Prathepha 2003; Prathepha and Baimai 2004) rice cultivars. Ayres et al. 1997 reported that $(CT)_{18}$ class had significantly lower levels of apparent amylose than all but one of the other classes, whereas the $(CT)_{11}$ class had significantly higher levels of amylose. In Thai rice, CT class with 11 repeats corresponds with high and intermediate amylose content (Prathepha 2003; Prathepha and Baimai 2004), while 18 repeat class corresponds with glutinous rice (Prathepha 2003).

In the present study, it was observed that majority of cultivars with allele 2 exhibited lower values for amylose content. However, four cultivars, viz., Ambemohar-Tambda, Kali kumud, Geerige sanna, and Vasane sanna bhatta, recorded high amylose content. A recent study by Dobo et al. (2010) reported that besides leader intron 5′ splice site, single-nucleotide polymorphisms in the exon 6 (A→C) and exon 10 (C→T) affect amylose content in US and European rice. Mikami et al. (2008) reported that five putative alleles at *Wx* locus are responsible for the continuous variation of amylose content in endosperm. Though *Wx* region is reported as a major QTL for amylose content (He et al. 1999; Tan et al. 1999; Lanceras et al. 2000; Septiningsih et al. 2003; Tian et al. 2005; Fan et al. 2005; Wang et al. 2007), epistasis, cytoplasmic effects, and the triploid nature of endosperm also complicate the inheritance of apparent amylose content Pooni et al. (1993). A study by Ayres et al. (1997) reported that the *Wx* locus alone was not enough to explain all the observed variation in apparent amylose content among rice cultivars and suggested possibility of involvement of some minor genes. In 2009, Tian et al. (2009) studied allelic diversities in rice starch biosynthesis resulting in a diverse array of rice eating and cooking qualities and recorded the presence of five genes that affect amylose content additively with *Wx*. The association of few high apparent amylose content rice with allele 2 recorded in the present study might be due to mutation in other parts of *Wx* locus or due to role of genes other than *Wx* in starch synthesis.

Despite the complex system of starch synthesis affecting amylose content in rice, markers linked with *Wx* locus have been successfully used for MAS (Liu et al. 2006; Yi et al. 2009, and Jin et al. 2010). Liu et al. (2006) developed CAPS marker for amylose content on the basis of single nucleotide polymorphism (G or T) located at the first nucleotide of the splice donor site of *Wx* intron 1 and found it useful in MAS for improving cooking and eating quality of rice hybrids. Two SSR markers RM 00 and RM 208 linked with *Wx* locus were used by Yi et al. (2009) for marker-assisted backcross breeding to improve cooking quality traits in Myanmar rice cultivar. A recent study by Jin et al. (2010) microsatellite allele $(CT)_{17}$ of *Wx* was utilized for hybrid breeding in Chinese rice.

The present study provides a valuable insight with respect to variation of *Wx* locus in Indian nonbasmati scented rice cultivars. Several scented rice varieties, including basmati and nonbasmati indigenous cultivars, contribute to Indian germplasm. Among these, scented rice fetches premium price owing to its aroma. The present study revealed that the SSR marker linked with *Wx* can segregate Indian rice cultivars, especially scented rice cultivars, on the basis of apparent amylose content. Thus, MAS for apparent amylose content – one of the important quality traits among scented rice – will open a new avenue in the breeding of quality rice.

5.4.2 Validation of RM 431 for Grain Length

The 54 Indian rice cultivars were subjected to PCR amplification of SSR RM 431 linked with QTL for grain length (Amarawathi et al. 2008). Separation on PCR products on native PAGE revealed the presence of four alleles. Among the cultivars assessed, alleles 1, 2, 3, and 4 were amplified in 14, 16, 3, and 17 cultivars respectively (Fig. 5.5). Four cultivars – Girga, Kaligajvili, Jaya, and Kernal local – were heterozygous with respect to this locus and amplified alleles 2 and 4 (indicated by black arrows in Fig. 5.4). The size of product varied from ~240 to ~260 bp as seen from Fig. 5.4. RM 431 located on chromosome 1 has (AG) repeat motif with 16 repeats and expected size of 251 bp (Temnykh et al. 2001, International Rice Genome Sequencing Project 2005). A study by Vikram et al. (2011) recorded polymorphism in RM 431 with product size of ~250 bp. Li et al. (2012) recorded seven alleles for RM 431 ranging from 239 to 255 bp. The size of product recorded in the present study is in agreement with these reports.

Presence of all four alleles of RM 431 was recorded in nonbasmati scented rice cultivars. Allele 1 was absent in nonscented rice cultivars and was present among scented rice cultivars of both basmati and nonbasmati types. Alleles 2 and 4 were amplified in nonscented rice cultivars. Among the cultivars assessed, allele 3 was rare and was amplified in nonbasmati scented rice cultivars Champakali, Velchi, and Bansaphool-A. Champakali and Velchi are landraces grown in Maharashtra state, India. Presence of rare allele in these cultivars indicated genetic variability and thus the potential of these cultivars to broaden the gene pool of scented rice cultivars.

Variation was exhibited in ambemohar types (Ambemohar, Ambemohar Ajra, Ambemohar Pandhara, Ambemohar-Tambda, and Ambemohar-157) with respect to the allele amplified for RM 431. Allele 2 was present in Ambemohar Pandhara and Ambemohar 157 and allele 4 in Ambemohar, Ambemohar Ajra and Ambemohar-Tambda. Thus, RM 431 allowed further segregation among Ambemohar cultivars.

Average polymorphism information content (PIC) of cultivars varied as follows: 0.354 in nonbasmati scented rice, 0.427 in basmati, and 0.22 in nonscented rice cultivars.

RM 431 is linked with QTL for grain length (Amarawathi et al. 2008). However, in the present study, no correlation is observed between allele sizes of RM 431 and grain length. Thus, RM 431 is not suitable to assess for grain length as a single marker in cultivars assed in present the study. RM 431 is linked with QTL for grain yield and is suitable for MAS (Vikram et al. 2011). RM 431 is also associated with plant height (Li et al. 2012). Considering the allelic variation of RM 431 recorded in the present study among Indian rice cultivars, RM 431 can be further used in the study of other linked QTLs among the set of cultivars assessed.

Grain length is mostly controlled by the major QTLs on chromosome 3 (Kubo et al. 2001; Xing et al. 2001; Li et al. 2004, Wan et al. 2005, 2006). Recent studies reveal that the markers developed from *GS3* can be used as effective functional markers for grain length (Fan et al. 2009; Ramkumar

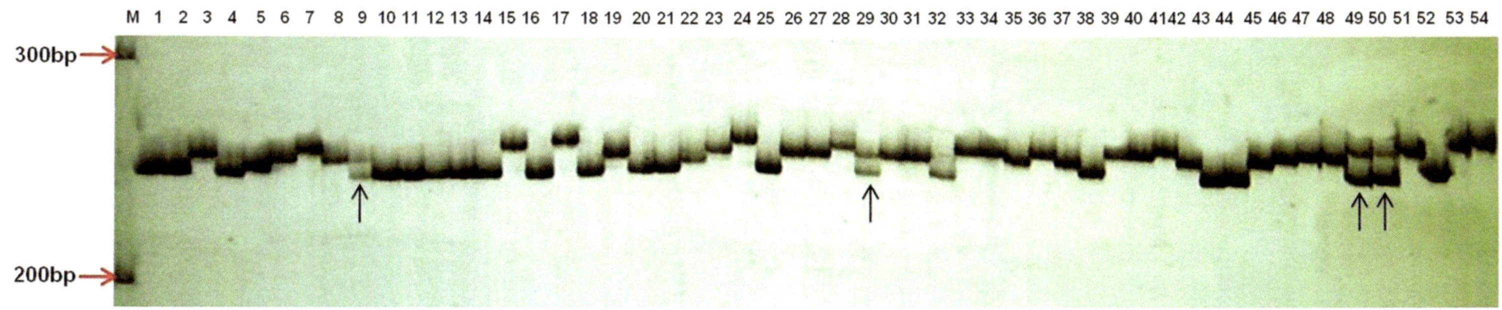

Lane M:100 bp ladder, 1:Ambemohar, 2:Ambemohar Ajra, 3:Ambemohar Pandhara, 4:Ambemohar-Tambda, 5:Champakali, 6:Chimansal, 7:Gham, 8:Ghansal, 9:Girga, 10:Jiri, 11:Kala bhat, 12:Kalsal, 13:Kamavatya,14: Kamod, 15:Kate chinoor, 16:Khadkya, 17:Kondhekar chinoor, 18:Kothmirsal, 19:Lal dodki, 20:Raibhog, 21:Tamsal, 22:Velchi, 23:Ambemohar-157, 24:Indrayani, 25:Basumati, 26:Gandhasale, 27:Geerige sanna, 28:Kagesali, 29:Kaligajvili, 30:Kali kumud, 31:Kumud, 32:Medhini sanna bhata, 33:Mysore sanna, 34:Sanna bili bhata, 35:Vasane sanna bhata, 36:Acharmati, 37:Badsahbhog, 38:Bansaphool-A, 39:Barke bhat, 40:Dubraj, 41:Dusara, 42:Girija samba, 43:Kalanamak, 44:Shyamjeer, 45:Kalakrishna, 46:Kothmbiri, 47:Kolamb, 48:Manila, 49:Jaya, 50:Kernal local, 51:Pakistan basmati, 52:Basmati 370, 53:Basmati 6311, 54:Pusa basmati-1

Fig. 5.5 *PAGE gel indicating PCR amplified products by RM 431 SSR primers among 54 rice cultivars. Black arrow* indicates heterozygous bands (Source: Mathure et al. 2014). 1: Ambemohar, 2: Ambemohar Ajra, 3: Ambemohar Pandhara, 4: Ambemohar-Tambda, 5: Champakali, 6: Chimansal, 7: Gham, 8: Ghansal, 9: Girga, 10: Jiri, 11: Kala bhat, 12: Kalsal, 13: Kamavatya, 14: Kamod, 15: Kate chinoor, 16: Khadkya, 17: Kondhekar chinoor, 18: Kothmirsal, 19: Lal dodki, 20: Raibhog, 21: Tamsal, 22: Velchi, 23: Ambemohar-157, 24: Indrayani, 25: Basumati, 26: Gandhasale, 27: Geerige sanna, 28: Kagesali, 29: Kaligajvili, 30: Kali kumud, 31: Kumud, 32: Medhini sanna bhata, 33: Mysore sanna, 34: Sanna bili bhata, 35: Vasane sanna bhata, 36: Acharmati, 37: Badsahbhog, 38: Bansaphool-A, 39: Barke bhat, 40: Dubraj, 41: Dusara, 42: Girija samba, 43: Kalanamak, 44: Shyamjeer, 45: Kalakrishna, 46: Kothmbiri, 47: Kolamb, 48: Manila, 49: Jaya, 50: Kernal local, 51: Pakistan basmati, 52: Basmati 370, 53: Basmati 6311, 54: Pusa basmati-1

et al. 2010; Wang et al. 2011). Fan et al. (2009) recorded that C-A polymorphism in the *GS3* gene is highly associated with grain length and explained grain length variation from 34.7 to 79.1 % in subpopulations. They (Fan et al. 2009) developed a cleaved amplified polymorphic sequence (CAPS) marker, SF28 as a functional marker based on C-A polymorphism which can identify long among medium and small grains. Recently, Ramkumar et al. (2010) have developed a PCR-based SNP marker system named DRR-GL, which targets the functional nucleotide polymorphism at *GS3* for selection of kernel length. Ramkumar et al. (2010) found DRR-GL as a suitable marker to identify all the basmati samples as long grain ones and recorded its limitation as marker in some nonbasmati samples. Wang et al. (2011) recorded polymorphic simple sequence repeats $(AT)_{(n)}$ at RGS1 and $(TCC)_{(n)}$ located in the last intron and the final exon of *GS3*. RGS1 and RGS2 were mainly associated with medium to short grain in Chinese rice and explained 26.4 and 26.2 % of the variation in grain length (Wang et al. 2011).

In the studies involving Pusa 1121 and Pusa 1342 by Amarawathi et al. (2008) and Singh et al. (2012), QTLs on chromosomes 1 and 7 were identified; however, QTL for the grain length gene *GS3* on chromosome 3 was not detected. The present study revealed that the RM 431 linked with QTL on chromosome 1 may be unique to Pusa 1121 as it was not sufficient to explain variation in grain length among the cultivars assessed. However, the cultivars can be subjected to *GS3*-based functional markers SF28, DRR-GL, RGS1, and RGS2 to assess utility of these markers in Indian cultivars.

The present study indicates the diversity of Indian cultivars with respect to *badh2*, *Wx*-SSR, and RM 431. This study highlights the potential of these cultivars to contribute to the gene pool owing to the presence of unique and rare alleles among scented rice landraces. Validation of markers for important quality traits such as amylose content, as revealed in the present study, can create enormous opportunities for the development of quality rice by MAS in scented rice of India.

References

Adair CR, Brachell HM, Jodon NE, Johnston TH, Thysell JR, Green VE, Webb BD, Atkins JG (1966) Rice breeding and testing methods in the U.S. USDA agricultural research service handbook, 289, pp 19–64

Aggarwal R, Shenoy V, Ramadevi J, Rajkumar R, Singh L (2002) Molecular characterization of some Indian Basmati and other elite rice genotypes using fluorescent-AFLP. Theor Appl Genet 105(5):680–690

Ahn SN, Bollisch CN, Tanksley SD (1992) RFLP tagging of a gene for aroma in rice. Theor Appl Genet 84:825–828

Aluko G, Martinez C, Tohme J, Castano C, Bergman C, Oard JH (2004) QTL mapping of grain quality traits from the interspecific cross *Oryza sativa*×*O. glaberrima*. Theor Appl Genet 109(3):630–639

Amarawathi Y, Singh R, Singh AK, Singh VP, Mohapatra T, Sharma TR, Singh NK (2008) Mapping of quantitative trait loci for basmati quality traits in rice *Oryza sativa* L. Mol Breed 21(1):49–65

Ayres NM, McClung AM, Larkin PD, Bligh HFJ, Jones CA, Park WD (1997) Microsatellites and a single nucleotide polymorphism differentiate apparent amylose classes in an extended pedigree of US rice germplasm. Theor Appl Genet 94:773–781

Bao JS, Corke H, Sun M (2002) Microsatellites in starch-synthesizing genes in relation to starch physicochemical properties in waxy rice *Oryza sativa L.* Theor Appl Genet 105:898–905

Bao JS, Corke H, Sun M (2006) Microsatellites single nucleotide polymorphisms and a sequence tagged site in starch-synthesizing genes in relation to starch physicochemical properties in nonwaxy rice *Oryza sativa* L. Theor Appl Genet 113(7):1185–1196

Bergman CJ, Delgado JT, McClung AM, Fjellstrom RG (2001) An improved method for using a microsatellite in the rice waxy gene to determine amylose class. Cereal Chem 78:257–260

Bhuyan N, Borah BK, Sarma RN (2007) Genetic diversity analysis in traditional lowland rice *Oryza sativa* L. of Assam using RAPD and ISSR markers. Curr Sci 93(7):967–972

Bligh HFJ, Till RI, Jones CA (1995) A microsatellite sequence closely linked to the wx gene of *Oryza sativa*. Euphytica 86:83–85

Bligh HFJ, Blackhall NW, Edwards KJ, McClung AM (1999) USING amplified fragment length polymorphisms and simple sequence length polymorphisms to identify cultivars of brown and white milled rice. Crop Sci 39:1715–1721

Bourgis F, Guyot R, Gherbi H, Tailliez E, Amabile I, Salse J, Ghesquiere A (2008) Characterization of the major fragrance gene from an aromatic japonica rice and analysis of its diversity in Asian cultivated rice. Theor Appl Genet 117(3):353–368

Bradbury LM, Fitgerald TL, Henry RJ, Jin Q, Waters DLE (2005a) The gene for fragrance in rice. Plant Biotech J 3:363–370

Bradbury LMT, Henry RJ, Jin Q, Reinke RF, Waters DLE (2005b) A perfect marker for fragrance genotyping in rice. Mol Breed 16:279–283

Chen S, Wu J, Yang Y, Shi W, Xu M (2006) The fgr gene responsible for rice fragrance was restricted within 69 kb. Plant Sci 171(4):505–514

Collard BC, Jahufer MZ, Brouwer JB, Pang EC (2005) An introduction to markers, quantitative trait loci (QTL) mapping and marker-assisted selection for crop improvement: The basic concepts. Euphytica 142(1):169–196

Cordeiro GM, Christopher MJ, Henry RJ, Reineke RF (2002) Identification of microsatellite markers for fragrance in rice by analysis of the rice genome sequence. Mol Breed 9:245–250

Das SR, Roy JK, Kar M, Das S (2003) Aromatic rices of Orissa. In: Singh RK, Singh US (eds) A treatise on the scented rices of India. Kalyani Publishers, New Delhi, pp 355–375

Davierwala AP, Chawdari KV, Kumar S, Reddy APK, Ranjekar PK, Gupta VS (2000) Use of three different marker systems to estimate genetic diversity of Indian elite rice varieties. Genetica 108:269–284

Deepa VL, Bhagyalakshmi N, Shashidhar HE, Singh V, Naidu AK (2009) Physicochemical and genetic analysis of an endemic rice variety, Njavara (*Oryza sativa* L.), in comparison to two popular South Indian cultivars, Jyothi (PTB 39) and IR 64G. J Agric Food Chem 57(24):11476–11483

Dobo M, Ayres N, Walker G, Park WD (2010) Polymorphism in the GBSS gene affects amylose content in US and European rice germplasm. J Cereal Sci 52(3):450–456

Fan CC, Yu XQ, Xing YZ, Xu CG, Luo LJ, Zhang QF (2005) The main effects, epistatic effects and environmental interactions of QTLs on the cooking and eating quality of rice in a doubled haploid line population. Theor Appl Genet 110:1445–1452

Fan C, Xing Y, Mao H, Lu T, Han B, Xu C, Li X, Zhang Q (2006) GS3, a major QTL for grain length and weight and minor QTL for grain width and thickness in rice, encodes a putative transmembrane protein. Theor Appl Genet 112(6):1164–1171

Fan CC, Yu SB, Wang CR, Xing YZ (2009) A causal C-A mutation in the second exon of GS3 highly associated with rice grain length and validated as a functional marker. Theor Appl Genet 118:465–472

Fitzgerald MA, Sackville Hamilton NR, Calingacion MN, Verhoeven HA, Butardo VM (2008) Is there a second fragrance gene in rice? Plant Biotechnol J 6(4):416–423

Fuentes JL, Escobar F, Alvarez A, Gallego G, Duque MC, Ferrer M, Deus JE, Tohme JM (1999) Analyses of genetic diversity in Cuban rice varieties using isozyme, RAPD and AFLP markers. Euphytica 109(2):107–115

Garland S, Lewin L, Blakeney A, Reineke R, Henry R (2000) PCR based molecular markers for the fragrance gene in rice (*Oryza sativa* L.). Theor Appl Genet 101:364–371

Ge XJ, Xing YZ, Xu CG, He YQ (2005) QTL analysis of cooked rice grain elongation, volume expansion, and water absorption using a recombinant inbred population. Plant Breed 124:121–126

He P, Li SG, Qian Q, Ma YQ, Li JZ, Wang WM, Chen Y, Zhu LH (1999) Genetic analysis of rice grain quality. Theor Appl Genet 98:502–508

He Y, Han Y, Jiang L, Xu C, Lu J, Xu M (2006) Functional analysis of starch-synthesis genes in determining rice eating and cooking qualities. Mol Breed 18:277–290

Huang N, Parco A, Mew T, Magpantay G, McCouch S, Guiderdoni E, Khush GS (1997) RFLP mapping of isozymes, RAPD and QTLs for grain shape, brown planthopper resistance in a doubled haploid rice population. Mol Breed 3(2):105–113

International Rice Genome Sequencing Project (IRGSP) (2005) The map-based of the rice genome. Nature 436:793–800

Jain S, Jain RK, McCouch SR (2004) Genetic analysis of Indian aromatic and quality rice (*Oryza sativa* L.) germplasm using panels of fluorescently-labeled microsatellite markers. Theor Appl Genet 109:965–977

Jena KK, Mackill DJ (2008) Molecular markers and their use in marker-assisted selection in rice. Crop Sci 48(4):1266–1276

Jin Q, Waters D, Cordeiro GM, Henry RJ, Reineke RF (2003) A single nucleotide polymorphism (SNP) marker linked to the fragrance gene in rice (*Oryza sativa* L.). Plant Sci 165:359–364

Jin L, Lu Y, Shao Y, Zhang G, Xiao P, Shen S, Bao J (2010) Molecular marker assisted selection for improvement of the eating, cooking and sensory quality of rice (*Oryza sativa* L). J Cereal Sci 51(1):159–164

Joshi RK, Behera L (2006) Identification and differentiation of indigenous non-basmati aromatic rice genotypes of India using microsatellite markers. Afr J Biotechnol 6(4):348–354

Joshi SP, Gupta VS, Aggarwal RK, Ranjekar PK, Brar DS (2000) Genetic diversity and phylogenetic relationship as revealed by inter simple sequence repeat (ISSR) polymorphism in the genus Oryza. Theor Appl Genet 100(8):1311–1320

Juliano BO (1985). Criteria and test for rice grain quality. In: Juliano BO (ed) Rice chemistry and technology. American Association of Cereal Chemists, Inc. St. Paul, Minnesota, USA, pp 443–513

Juliano BO, Duff B (1991) Rice grain marketing and quality issues. In: Rice grain quality as an emerging priority in national breeding programs. International Rice Research Institute, Los Banos, pp 55–64

Kovach MJ, Calingacion MN, Fitzgerald MA, McCouch SR (2009) The origin and evolution of fragrance in rice (*Oryza sativa* L.). Proc Natl Acad Sci U S A 106:14444–14449

Kubo T, Takano-kai N, Yoshimura A (2001) RFLP mapping of genes for long kernel and awn on chromosome 3 in rice. Rice Genet Newsl 18:26–28

Kumar LS (1999) DNA markers in plant improvement: an overview. Biotechnol Adv 17(2–3):143–182

Lanceras JC, Hun ZL, Naivikul Q, Vanavichit A, Ruanjaichon V, Tragoonrung S (2000) Mapping of genes for cooking and eating qualities in Thai Jasmine rice (KDML105). DNA Res 7:93–101

Langridge P, Lagudah E, Holton T, Appels R, Sharp P, Chalmers K (2001) Trends in genetic and genome analyses in wheat: a review. Aust J Agric Res 52:1043–1077

Larkin PD, Park WD (2003) Association of waxy gene single nucleotide polymorphisms with starch characteristics in rice (*Oryza sativa* L.). Mol Breed 12:335–339

Li J, Thomson M, McCouch SR (2004) Fine mapping of a grain-weight quantitative trait locus in the pericentromeric region of rice chromosome 3. Genetics 168(4):2187–2195

Li X, Yan W, Agrama H, Jia L, Jackson A, Moldenhauer K, Yeater K, McClung A, Wu D (2012) Unraveling the complex trait of harvest index with association mapping in rice (*Oryza sativa* L.). PLoS One 7(1):e29350

Lin H (1995) RFLP mapping of QTLs for grain shape traits in indica rice (*Oryza sativa* L. subsp. *indica*). Sci Agric Sin 28:1–7

Liu X, Gu M, Han Y, Ji Q, Lu J, Gu S, Zhang R, Li X, Chen J, Korban SS, Xu M (2004) Developing gene-tagged molecular markers for functional analysis of starch-synthesizing genes in rice (*Oryza sativa* L.). Euphytica 135(3):345–353

Liu QQ, Li QF, Cai XL, Wang HM, Tang SZ, Yu HX, Wang ZY, Gu MH (2006) Molecular marker-assisted selection for improved cooking and eating quality of two elite parents of hybrid rice. Crop Sci 46 2354–2360

Lorieux M, Petrov M, Huang N, Guiderdoni E, Ghesquière A (1996) Aroma in rice: genetic analysis of a quantitative trait. Theor Appl Genet 93:1145–1151

Mackill DJ, Zhang Z, Redona ED, Colowit PM (1996) Level of polymorphism and genetic mapping of AFLP markers in rice. Genome 39:969–977

Mathure S, Nadaf A, Jawali N (2010) Diversity analysis in selected non-basmati scented rice collection (*Oryza sativa* L.). Rice Sci 17(1):35–42

Mathure S, Shaikh A, Renuka N, Wakte K, Jawali N, Thengane R, Nadaf A (2011) Characterisation of aromatic rice (*Oryza sativa* L.) germplasm and correlation between their agronomic and quality traits. Euphytica 179(2):237–246

Mathure SV, Jawali N, Thengane RJ, Nadaf AB (2014) Comparative quantitative analysis of headspace volatiles and their association with BADH2 marker in non-basmati scented, basmati and non-scented rice (Oryza sativa L.) cultivars of India. Food Chem 142:383–391

McGregor C, Lambert C, Geryling M, Louw J, Warnich L (2000) A comparative assessment of DNA fingerprinting techniques (RAPD, ISSR, AFLP and SSR) in tetraploid potato (*Solanum tuberosum*) germplasm. Euphytica 13:135–144

McKenzie KS, Rutger JN (1983) Genetic analysis of amylose content, alkali spreading score, and grain dimensions in rice. Crop Sci 23:306–313

Mikami I, Uwatoko N, Ikeda Y, Yamaguchi J, Hirano HY, Suzuki Y, Sano Y (2008) Allelic diversification at the wx locus in landraces of Asian rice. Theor Appl Genet 116:979–989

Mondini L, Arshiya N, Pagnotta MA (2009) Assessing plant genetic diversity by molecular tools. Diversity 1:19–35

Mueller U, Wolfenbarger L (1999) AFLP genotyping and fingerprinting. Tree 14:389–394

Myint KM, Arikit S, Wanchana S, Yoshihashi T, Choowongkomon K, Vanavichit A (2012). A PCR-based marker for a locus conferring the aroma in Myanmar rice (*Oryza sativa* L.) Theor Appl Genet 1–10

Nagaraju J, Kathirvel M, Kumar RR, Siddiq EA, Hasnain SE (2002) Genetic analysis of traditional and evolved Basmati and non-Basmati rice varieties by using fluorescence-based ISSR-PCR and SSR markers. Proc Natl Acad Sci U S A 99(9):5836–5841

Nei M, Li WH (1979) Mathematical Model for studying genetic variation in terms of Restriction endonucleases. PNAS USA 76:5269–5273

Ovesna J, Polakova K, Lisova L (2002) DNA analyses and their applications in plant breeding. Czech J Genet Plant Breed 38:29–40

Pooni HS, Kumar I, Khush GS (1993) Genetical control of amylose content in selected crosses of indica rice. Heredity 70:269–280

Prashanth SR, Parani M, Mohanty BP, Talame V, Tuberosa R, Parida A (2002) Genetic diversity in cultivars and landraces of *Oryza sativa* subsp. indica as revealed by AFLP markers. Genome 45(3):451–459

Prathepha P (2003) Characterization of Waxy microsatellite classes that are closely linked to the rice Waxy gene and amylose content in Thai rice germplasm. Songklanakarin J Sci Technol 25:1–8

Prathepha P, Baimai V (2004) Variation of Wx microsatellite allele, waxy allele distribution and differentiation of chloroplast DNA in a collection of Thai rice (*Oryza sativa* L.). Euphytica 140:231–237

Raghunathachari P, Khanna VK, Singh US, Singh NK (2000) RAPD analysis of genetic variability in Indian scented rice germplasm (*Oryza sativa* L.). Curr Sci 79(7):994–998

Rahemi MR, Kazemitabar SK, Moumeni A, Ebadi AA (2007) Evaluation of genetic diversity by using link marker for amylose content of some Iranian local rice cultivars. Asian J Plant Sci 6(5):874–877

Ramkumar G, Sivaranjani AK, Pandey MK, Sakthivel K, Shobha Rani N, Sudarshan I, Prasad GSV, Neeraja CN, Sundaram RM, Viraktamath BC, Madhav MX (2010) Development of a PCR-based SNP marker system for effective selection of kernel length and kernel elongation in rice. Mol Breed 26(4):735–740

Rana RS (1995) Conservation of plant genetic resources in India. In: Rana RS, Chandel KPS, Bhat SR, Mandal BB, Karihaloo JL, Bhat KV, Pandey R (eds) Plant germplasm conservation: biotechnological approaches. NBPGR, New Delhi

Rashid M, Ren-hu L, Wei J, Yong-han X, Fu-lin W, Yue-zhi T, Jun-mei W, Cheema AA, Jin-qing C, He G

(2009) Genomic diversity among Basmati rice (*Oryza sativa* L.) mutants obtained through ^{60}Co gamma radiations using AFLP markers. Afr J Biotechnol 8(24):6777–6783

Raychaudhari P, Kohli S, Srinivasan K, Mohapatra T, Sharma RP (2001) Identification and classification of aromatic rices based on DNA fingerprinting. Euphytica 118:243–251

Redona ED, Mackill DJ (1998) Quantitative trait locus analysis for rice panicle and grain characteristics. Theor Appl Genet 96(6):957–963

Rohilla R, Singh VP, Singh US, Singh RK, Khush GS (2000) Crop husbandry and environmental factors affecting aroma and other quality traits In: Singh RK, Singh US, Khush GS (eds) Aromatic rices. Oxford & IBH Pub. Co., New Delhi, pp 155–177

Roy S, Banerjee A, Senapati BK, Sarkar G (2012) Comparative analysis of agro-morphology, grain quality and aroma traits of traditional and Basmati-type genotypes of rice, *Oryza sativa* L. Plant Breed 131(4):486–492

Saini N, Jain N, Jain S, Jain RK (2004) Assessment of genetic diversity within and among Basmati and non-Basmati rice varieties using AFLP, ISSR and SSR markers. Euphytica 140:133–146

Saker MM, Youssef SS, Abdallah NA, Bashandy HS, El Sharkawy AM (2005) Genetic analysis of some Egyptian rice genotypes using RAPD, SSR and AFLP. Afr J Biotechnol 4(9):882–890

Sakthivel K, Sundaram RM, Shobharani N, Balachandran SM, Neeraja CN (2009) Genetic and molecular basis of fragrance in rice. Biotechnol Adv 27(4):468–473

Semagn K, Bjørnstad A, Ndjiondjop MN (2006) An overview of molecular marker methods for plants. Afr J Biotechnol 5(25):2540–2568

Septiningsih EM, Trijatmiko KR, Moeljopawiro S, McCouch SR (2003) Identification of quantitative trait loci for grain quality in an advanced backcross population derived from the *Oryza sativa* variety IR64 and the wild relative *O. rufipogon*. Theor Appl Genet 107:1433–1441

Shao G, Tang S, Chen M, Wei X, He J, Luo J, Jiao G, Hu Y, Xie L, Hu P (2013) Haplotype variation at Badh2, the gene determining fragrance in rice. Genomics 101:157–162

Sharma SD, Tripathy S, Biswal J (2000) Origin of Oryza sativa and its ecotypes. Rice Breed. Genet. Edited by Oxford and IBH Pub. Co., New Delhi, pp. 350–369

Shi W, Yang Y, Chen S, Xu M (2008) Discovery of a new fragrance allele and the development of functional markers for the breeding of fragrant rice varieties. Mol Breed 22(2):185–192

Shobharani N, Singh RK (2003) Efforts on aromatic rice improvement in India. In: Singh RK, Singh US (eds) A treatise on the scented rices of India. Kalyani Publishers, New Delhi, pp 23–72

Shu QY, Wu DX, Xia YW, Gao MW, Ayres NM, Larkin PD, Park WD (1999) Microsatellites polymorphism on the waxy gene locus and their relationship to amylose content in indica and japonica rice, *Oryza sativa* L. Acta Genet Sin 26:350–358

Shure M, Wessler S, Fedoroff N (1983) Molecular identification and isolation of the Waxy locus in maize. Cell 35:225–233

Siddiq EA, Vemireddy LR, Nagaraju J (2012) Basmati rices: genetics, breeding and trade. Agric Res 1(1):25–36

Singh RK, Singh US (2003) A treatise on the scented rices of India. Kalyani Publishers, New Delhi

Singh KN, Nandi R, Shanmugasundram P, Sadasivam S, Huang N, Brar DS, Khush GS (1999) High resolution DNA fingerprinting of Indian rice (*Oryza sativa* L.) varieties by amplified fragment length polymorphism. Genet Resour Crop Evol 46(5):427–433

Singh RK, Singh US, Khush GS, Rohilla R, Singh JP, Singh G, Shekhar KS (2000) Small and medium grained scented rices of India. In: Singh RK, Singh US, Khush GS (eds) Aromatic rices. Oxford & IBH Pub. Co, New Delhi, pp 155–177

Singh RK, Sharma RK, Singh AK, Singh VP, Singh NK, Tiwari SP, Mohapatra T (2004) Suitability of mapped sequence tagged microsatellite site markers for establishing distinctness, uniformity and stability in aromatic rice. Euphytica 135:135–143

Singh D, Sirohi A, Vikas SS, Kumar V, Gaur A, Chaudhari R, Chaudhari N (2006) RAPD based identification of farmer's collections of Kalanamak and traditional varieties of basmati rice. Ind J Crop Sci 1(1–2):102–105

Singh R, Singh AK, Sharma TR, Singh A, Singh NK (2012) Fine mapping of grain length QTLs on chromosomes 1 and 7 in Basmati rice (Oryza sativa L.). J Plant Biochem Biotechnol 21(2):157–166

Sivaranjani AKP, Pandey MK, Sudharshan I, Ramkumar G, Madhav MS, Sundaram RM, Varaprasad GS, Shobharani N (2010) Assessment of genetic diversity among Basmati and non-Basmati aromatic rices of India using SSR markers. Curr Sci 99(2):221–226

Smith AM, Denyer K, Martin C (1997) The synthesis of the starch granule. Annu Rev Plant Physiol Plant Mol Biol 48:67–87

Sood BC, Siddiq EA, Zaman FU (1983) Genetic analysis of kernel elongation in rice. Ind J Genet 43:40–43

Sun SX, Gao FY, Lu XJ, Wu XJ, Wang XD, Ren GJ, Luo H (2008) Genetic analysis and gene fine mapping of aroma in rice (Oryza sativa L. Cyperales, Poaceae). Genet Mol Biol 31(2):532–538

Tan YF, Zhang QF (2001) Correlation of simple sequence repeat (SSR) variants in the leader sequence of the wx gene with amylose content of the grain in rice. Acta Bot Sin 43:146–150

Tan YF, Li JX, Yu SB, Xing YZ, Xu CG, Zhang QF (1999) The three important traits for cooking and eating quality of rice grains are controlled by a single locus in an elite rice hybrid, Shanyou 63. Theor Appl Genet 99:642–648

Temnykh S, DeClerck G, Lukashova A, Lipovich L, Cartinhour S, McCouch S (2001) Computational and experimental analysis of microsatellites in rice (*Oryza sativa* L.): frequency, length variation, transposon associations, and genetic marker potential. Genome Res 11:1441–1452

Tian R, Jiang GH, Shen LH, Wang LQ, He YQ (2005) Mapping quantitative trait loci underlying the cooking and eating quality of rice using a DH population. Mol Breed 15:117–124

Tian ZX, Qian Q, Liu QQ, Yan MX, Liu XF, Yan CJ, Liu GF, Gao ZY, Tang SZ, Zeng DL, Wang YH, Yu JM, Gu MH, Li JY (2009) Allelic diversity in rice starch biosynthesis pathway leads to a diverse array of rice eating and cooking qualities. Proc Natl Acad Sci U S A 106:21760–21765

Vanavichit A (2007) Molecular diversity and evolution of aromatic rice in Thailand. In: International training workshop, the conservation and utilization of tropical/subtropical plant genetic resources. Taiwan Agricultural Research Institute, Council of Agriculture (TARIC), Taiwan, pp 131–134

Van de Peer Y, De Wachter R (1993) TREECON: a software package for the construction and drawing of evolutionary trees. Comput Appl Biosci 9:177–182

Vikram P, Swamy BP, Dixit S, Ahmed H, Cruz MT, Singh A, Kumar A (2011) qDTY1.1, a major QTL for rice grain yield under reproductive-stage drought stress with a consistent effect in multiple elite genetic backgrounds. BMC Genet 12(1):89

Vos P, Hogers R, Bleeker M, Reijians M, Lee T, Hornes M, Frijters A, Pot J, Peleman J, Kuiper M, Zabeau M (1995) AFLP, a new technique for DNA fingerprinting. Nucl Acids Res 23:4407–4414

Wan XY, Wan JM, Weng JF, Jiang L, Bi JC, Wang CM, Zhai HQ (2005) Stability of QTLs for rice grain dimension and endosperm chalkiness characteristics across eight environments. Theor Appl Genet 110(7):1334–1346

Wan XY, Wan JM, Jiang L, Wang JK, Zhai HQ, Weng JF, Wang HL, Lei CH, Wang JL, Zhang X, Cheng ZJ, Guo XP (2006) QTL analysis for rice grain length and fine mapping of an identified QTL with stable and major effects. Theor Appl Genet 112:1258–1270

Wanchana S, Kamolsukyunyong W, Ruengphayak S, Toojinda T, Tragoonrung S, Vanavichit A (2005) A rapid construction of a physical contig across a 4.5 cM region for rice grain aroma facilitates marker enrichment for positional cloning. Sci Asia 31:299–306

Wang Z, Wu Z, Xing Y, Zheng F, Gao J, Hong M (1990) Nucleotide sequence of rice waxy gene. Nucleic Acids Res 18:5898

Wang ZY, Zhang FQ, Shen GZ, Gao JP, Snustad DP, Li MG, Zhang JL, Hong MM (1995) The amylase content in rice endosperm is related to the post-transcriptional regulation of the waxy gene. Plant J 7:613–622

Wang J, Wan X, Crossa J, Crouch J, Weng J, Zhai H, Wan J (2006) QTL mapping of grain length in rice (*Oryza sativa* L.) using chromosome segment substitution lines. Genet Res 88(2):93–104

Wang LQ, Liu WJ, Xu Y, He YQ, Luo LJ, Xing YZ, Xu CG, Zhang Q (2007) Genetic basis of 17 traits and viscosity parameters characterizing the eating and cooking quality of rice grain. Theor Appl Genet 115:463–476

Wang C, Chen S, Yu S (2011) Functional markers developed from multiple loci in GS3 for fine marker-assisted selection of grain length in rice. Theor Appl Genet 122(5):905–913

Xing YZ, Tan YF, Xu CG, Hua JP, Sun XL (2001) Mapping quantitative trait loci for grain appearance traits of rice using a recombinant inbred line population. Acta Bot Sin 43:721–726

Xu Y, Crouch JH (2008) Marker-assisted selection in plant breeding: from publications to practice. Crop Sci 48:391–407

Yi M, Nwe KT, Vanavichit A, Chai-arree W, Toojinda T (2009) Marker assisted backcross breeding to improve cooking quality traits in Myanmar rice cultivar Manawthukha. Field Crop Res 113(2):178–186

Zhu J, Gale MD, Quarrie S, Jackson MT, Bryan GJ (1998) AFLP markers for the study of rice biodiversity. Theor Appl Genet 96(5):602–611

Index

A. Nadaf et al., *Scented rice (Oryza sativa L.) Cultivars of India: A Perspective on Quality and Diversity*, DOI 10.1007/978-81-322-2665-9